To Dr. R.G. Poulter

With compliments & best wishes

17.11.88

FOOD PRESERVATION BY MOISTURE CONTROL

Proceedings of an international symposium held in Penang, 21–24 September 1987, under the joint auspices of ISOPOW/IUFOST, The Malaysian Institute of Food Technology and Universiti Sains Malaysia.

FOOD PRESERVATION BY MOISTURE CONTROL

Edited by

C. C. SEOW

Food Technology Division, School of Industrial Technology, Universiti Sains Malaysia, Penang, Malaysia

Assistant Editors

T. T. TENG and C. H. QUAH

School of Industrial Technology, Universiti Sains Malaysia, Penang, Malaysia

ELSEVIER APPLIED SCIENCE
LONDON and NEW YORK

ELSEVIER APPLIED SCIENCE PUBLISHERS LTD
Crown House, Linton Road, Barking, Essex IG11 8JU, England

Sole Distributor in the USA and Canada
ELSEVIER SCIENCE PUBLISHING CO., INC.
52 Vanderbilt Avenue, New York, NY 10017, USA

WITH 59 TABLES AND 79 ILLUSTRATIONS

British Library Cataloguing in Publication Data

Food preservation by moisture control.
1. Food. Preservation
I. Seow, C. C.
641.4

ISBN 1-85166-261-8

Library of Congress CIP data applied for

Printed in Northern Ireland by The Universities Press (Belfast) Ltd.

Dedicated To

DR. R.B. DUCKWORTH

PREFACE

In 1974, the International Symposium on Water Relations of Foods, organised by Dr. R.B. Duckworth, was held at the University of Strathclyde, Glasgow, Scotland. That symposium was to be the forerunner of a series of international symposia on the properties of water in foods, widely known as ISOPOW or Duckworth meetings. ISOPOW II, III and IV have since been held in Japan (1978), France (1983) and Canada (1987), but never in the lesser developed tropical regions where losses of food materials due to spoilage are far more substantial and where examples also abound of traditional products relying for their keeping quality primarily on control of the internal aqueous environment. It was Dr. Duckworth who first mooted the idea of holding a supplementary ISOPOW meeting in tropical Asia to examine the application of scientific knowledge on the behaviour of water in foods towards the more effective preservation of food materials in the tropics. The International Symposium on Preservation of Foods in Tropical Regions by Control of the Internal Aqueous Environment held in Penang, Malaysia from 21 - 24 September, 1987 fulfilled such an aim by bringing together some of the world's foremost authorities on the properties of water in foods and scientific workers immediately concerned with food preservation in the tropics, especially through the application of low and intermediate moisture food technology. Special tribute is paid to Dr. Duckworth, without whose vision, invaluable advice and expert guidance, plans for this Symposium would not have been brought to fruition.

This volume is based on selected papers presented at the Symposium. It deals firstly with the fundamental aspects of the state of water in foods, followed by the more practical aspects of food preservation by control of water activity or moisture content as applied to different food commodities.

C.C. SEOW

ACKNOWLEDGEMENTS

The Symposium was jointly organised by the Malaysian Institute of Food Technology (MIFT), the ISOPOW/IUFoST Executive Committee and Universiti Sains Malaysia (USM). The Symposium Organisers gratefully acknowledge the financial support from the following : Australian International Development Assistance Bureau (AIDAB), Commonwealth Foundation, Kuok Foundation, Malaysian Freeze-Drying Corporation Sdn. Bhd., Malpom Industries Bhd., Nutritional Products Sdn. Bhd., Antah Sri Radin Sdn. Bhd., Bayer (M) Sdn. Bhd., Glaschem Laboratory Supplier Sdn. Bhd., Medilab Sales & Service, Siber Hegner (M) Sdn. Bhd., Advance Scientific &

Medical Supplies Sdn. Bhd., Serba Olah Sdn. Bhd., Syarikat Quicklab Enterprise, Premier Milk (M) Sdn. Bhd., Century Chemicals Sdn. Bhd., Hisco (M) Sdn. Bhd., Hewlett Packard, Chocolate Products (M) Sdn. Bhd., Ajinomoto (M) Bhd, United Malaysian Flour Mills Bhd., Novo Industries A/S, Fraser & Neave Bhd., and General Scientific Co. Sdn. Bhd. The Organisers are also indebted to Malaysia Airlines, Tourist Development Corporation of Malaysia and Orchid Hotel for assistance rendered.

The editor would like to record his thanks especially to Yang Berbahagia Datuk Hj. Musa Mohamad, Vice-Chancellor of USM; Mr. J.F. Kefford, Secretary-General of IUFoST; Dr. R.B. Duckworth, President of the ISOPOW Executive Committee; Mr. S.A. Goh, Secretary of MIFT; Members of the International Advisory Committee and Local Coordinating Committee; Session Chairmen, contributors of papers and all participants; Dr. T.T. Teng and Mr. C.H. Quah, the assistant editors; Ms. Rahmah Puteh and Mrs. Hanim Noor for typing and secretarial services; and all other co-workers who helped in one way or another to make the Symposium a success.

FOREWORD

I feel honoured to have been asked by Dr. Seow to write a foreword for the proceedings of a meeting which I regard as having broken important new ground in a number of different respects. This was the first of what it is hoped will be a new series of meetings sponsored jointly by ISOPOW (International Symposia on the Properties of Water), together with other international and national bodies concerned in the food sciences, to address specific practical problems connected with the state of water in food materials.

The original concept of the meeting was formulated in 1983, after ISOPOW III in Beaune, and was finally successfully realised in September, 1987, after a great deal of hard work by Dr. Seow and his colleagues, based in the Universiti Sains Malaysia.

The plan involved bringing together leading international authorities on chemical and biological aspects of the state of water in foods, and technologists and research workers who are concerned, on a day to day basis, with combatting the practical problems of extending the useful life of essential food supplies under tropical conditions.

Throughout man's history, certainly since the dawn of civilisation, the preservation of food materials has played a vital part in avoiding widespread starvation and it continues to do so to this day, even in the industrially developed world. Nowhere, however, is preservation by control of the condition of the water in the food more widely needed, while at the same time fraught with much difficulty, as in the tropical parts of the world.

The choice of subject for the first of this new series of meetings was therefore most appropriate and speakers were recruited who could speak with great authority both on fundamental and on practical topics.

The setting and timing of the meeting, in the week immediately prior to the holding of the International Congress of Food Science and Technology in Singapore, was also most fortunate, because it maintained a tradition under which, since the Madrid Congress of 1974, an ISOPOW meeting has been closely associated with each subsequent IUFoST Congress.

Important advances have been taking place recently in our knowledge concerning the condition of water in biological materials and the validity of methods used for characterising the state of water and these were brought out during the early sessions of the meeting. Later sessions dealt, on a commodity basis, with the results of practical studies on the application of preservative processes in various parts of the tropics, with particular reference to S.E. Asia. Finally, a summary of each session was presented by the respective session chairmen and the meeting ended with a general discussion.

A personal view of the lessons emerging during the course of this meeting, which I am sure all those attending found most instructive as well as enjoyable, may be summarised as follows:-

Those more immediately involved in practical aspects of methods of food preservation which depend for their effectiveness on control of the aqueous environment within a material must remain alive to the fact that newer research is currently leading to important changes in our understanding of the properties of water in foods, and that some previously widely-held views on the theoretical background to their activities are no longer tenable. At the same time, methods such as the measurement of relative water vapour pressure (a_w) which have long been used for characterising the state of water in foods and which continue to prove empirically highly useful, should still be exploited to their fullest advantage, yet with a greater appreciation of their theoretical limitations.

Progress in basic research will no doubt continue to necessitate further modifications to our background understanding. At the same time, much valuable empirical experience, often awaiting scientific explanation, resides in techniques of preservation which have already been applied in local areas over long periods of time, and this can both aid in the fight to reduce the extent of food spoilage in other regions and also provide additional models against which our scientific interpretations may be further tested.

Finally, therefore, Dr. Seow and his team are to be congratulated on an exceptionally well-organised, highly successful and most enjoyable symposium which, of its kind, will serve as an important milestone as well as a model on which future meetings in the series may be confidently based.

R.B. DUCKWORTH
President, ISOPOW Executive Committee

CONTENTS

Section 3. Food Commodities of Plant Origin

LIST OF CONTRIBUTORS

K.A. Buckle
Department of Food Science and Technology, University of New South Wales, P.O. Box 1, Kensington, NSW 2033, Australia.

C.L. Chong
Palm Oil Research Institute of Malaysia, No. 6 Persiaran Institusi, Bandar Baru Bangi, 43000 Kajang, Selangor, Malaysia.

J.H.B. Christian
CSIRO Division of Food Research, P.O. Box 52, North Ryde, NSW 2113, Austalia.

E.C. Chuah
Food Technology Division, Malaysian Agricultural Research and Development Institute, GPO Box No. 12301, 50774 Kuala Lumpur, Malaysia.

P.E. Doe
Department of Civil and Mechanical Engineering, University of Tasmania, GPO Box 252C, Hobart, Tasmania 7001, Australia.

G.W. Gould
Unilever Research Laboratory, Colworth House, Sharnbrook, Bedford, MK44 1LQ, England.

S. Guilbert
CIRAD/CEEMAT, Food Technology Programme, Domaine de Lavalette, Avenue du Val-de-Montferrand, 34100 Montpellier, France.

E.S. Heruwati
Research Institute for Fishery Technology, P.O. Box 30, Palmerah, Jakarta, Indonesia.

A.D. Hocking
CSIRO Division of Food Research, P.O. Box 52, North Ryde, NSW 2113, Australia.

A.B.H. Hussin
Food Technology Division, Malaysian Agricultural Research and Development Institute, GPO Box No. 12301, 50774 Kuala Lumpur, Malaysia.

K.S. Jayaraman
Defence Food Research Laboratory, Mysore 570011, India.

M. Karel
Department of Applied Biological Sciences, Massachusetts Institute of Technology, Cambridge, MA 02139, USA.

Z.A. Obanu
Department of Food Science and Technology, University of Nigeria, Nsukka, Nigeria.

A.S.H. Ong
Palm Oil Research Institute of Malaysia, No. 6 Persiaran Institusi, Bandar Baru Bangi, 43000 Kajang, Selangor, Malaysia.

H. Purnomo
Department of Animal Product Technology, University of Brawijawa, Malang, East Java, Indonesia.

S. Putro
Research Institute for Fishery Technology, P.O. Box 30, Palmerah, Jakarta, Indonesia.

S. Sastrodiantoro
Department of Food Science and Technology, University of Brawijawa, Malang, East Java, Indonesia.

C.C. Seow
Food Technology Division, School of Industrial Technology, Universiti Sains Malaysia, 11800 Penang, Malaysia.

D. Simatos
Universite de Dijon, Ecole Nationale Superieure de Biologie Appliquee a la Nutrition et a l'Alimentation, Laboratoire de Biologie Physico-Chimique, Campus Universitaire, 21100 Dijon, France.

R.A. Souness
Department of Food Science and Technology, University of New South Wales, P.O. Box 1, Kensington, NSW 2033, Australia.

R.J. Steele
CSIRO Division of Food Research, P.O. Box 52, North Ryde, NSW 2113, Australia.

K. Thevamalar
Food Technology Division, School of Industrial Technology, Universiti Sains Malaysia, 11800 Penang, Malaysia.

G. Vuataz
Nestle Research Center, Nestec Ltd, Vers-chez-les-Blanc, CH-1000 Lausanne 26, Switzerland.

P. Wuttijumnong
Department of Agro-Industry, Prince of Songkhla University, Haad Yai, Thailand.

Q.L. Yeoh
Food Technology Division, Malaysian Agricultural Research and Development Institute, GPO Box 12301, 50774 Kuala Lumpur, Malaysia.

CHARACTERIZATION OF THE CONDITION OF WATER IN FOODS-PHYSICO-CHEMICAL ASPECTS

D. SIMATOS
Universite de Dijon
Ecole Nationale Superieure de Biologie
Appliquee a la Nutrition et a l'Alimentation,
Laboratoire de Biologie Physico-Chimique,
Campus Universitaire
21000 Dijon, France

AND

M. KAREL
Department of Applied Biological Sciences,
Massachusetts Institute of Technology,
Cambridge, MA 02139, USA

INTRODUCTION

Water in foods is subject to interactions which change its properties and also change the properties of the components with which it interacts. The task of the food scientist is to develop methods for characterization and quantification of these changes, and to examine the possible consequences of these changes on food quality. The present paper is devoted to the review of these points in connection with the physico-chemical aspects of food stability.

PHYSICO-CHEMICAL PROPERTIES OF WATER IN FOODS

Vapour Pressure and Water Activity

The most commonly used and most effective physical method of assessing the state of water in food is the measurement of the partial vapour pressure of water in equilibrium with a given moisture content, at a constant temperature. The almost universally accepted convention is the use of the relative vapour pressure and its designation as "water activity".

$$a = \frac{P}{P_o} \qquad (1)$$

where

a = water activity
P = partial pressure of water in the food, at a given temperature
P_o = vapour pressure of water at the same temperature.

It is of course well known that the above definition of water activity is not necessarily equivalent to the thermodynamic definition based on the chemical potential:

$$\mu_w - \mu_w^o = RT\ln a_w \qquad (2)$$

where μ_w is chemical potential of water defined, in turn, as:

$$\mu_w = \left(\frac{\Delta G}{\Delta n_w}\right)_{T,P,n_i} \qquad (3)$$

where

G = Gibbs Free Energy
n = number of moles
T = temperature
P = pressure
Subscripts w and i refer to water and other components, respectively.
Superscript (o) refers to a defined "standard" state.

For practical purposes, at most conditions of significance to food scientists, the definition of Equation (1) adequately determines the key function of water activity, that of defining equilibrium between phases with respect to water. At constant temperature, any two phases such as air and food exposed to it; oil and droplets of aqueous solution within it; and a bacterial cell and the medium surrounding it are at equilibrium with respect to water when their water activities are equal. Furthermore, in most cases, the rates of transfer of water between phases are proportional to the difference in water activity between the phases.

There have been recent objections (Franks, 1985a) to the use of the term "water activity" in non-equilibrium situations. We are not persuaded by these objections. Non-equilibrium situations arise from two sources:

1. An existing moisture and therefore also a partial pressure gradient in one of the phases. In this situation, it is simply inappropriate to refer to any property of the phase as a whole. Each position within the phase has its own property :

$$a = f(x,y,z,t) \qquad (4)$$

where x,y, and z are the usual space coordinates and t is time.

2. An existing metastable situation with respect to other components of the phase. Examples include : crystallization of sugars, swelling of polymers, gelatinization of starch, denaturation of proteins, diffusion of humectants, and others. In this situation, we have again the non-equilibrium state defined by Equation (4), but it arises from changes in state or spacial distribution of components other than water. When equilibrium does not exist, non-equilibrium relations must be used.

The graphical or analytical representation of the relation between a and m (moisture content), so-called sorption isotherms, are readily determined, at least for samples for which an equilibrium or "pseudo-equilibrium" state can be established, and are an indispensible tool of the food technologist (Karel et al., 1975; Iglesias & Chirife 1982; Labuza, 1984; Van den Berg, 1986). At a minimum, they provide the necessary information for food formulation (i.e. the moisture contents of the different components which will be in equilibrium with each other); for packaging (required resistance to water transfer); and, as will be discussed elsewhere in this volume, for the expected range of potential microbial growth. They may also provide less definite, but nevertheless important information about the system. Thus the presence of hysteresis will point to conformational or phase changes in components, and quantities derived from the isotherms are useful in estimating the number and water affinity of specific binding sites, although the theoretical application of such data has to be made with some caution.

Thus the BET monolayer value obtained from Equation (5) is a good estimate of the number of primary sites for water sorption, and Equation (6) may be used to estimate the heat of sorption at any given extent of coverage, i.e. a given moisture content.

$$\frac{m}{m_1} = \frac{C\,a}{(1 - a)(1 + (C - 1)a)} \qquad (5)$$

$$\frac{d(\ln a)}{d(1/T)} = \frac{-\,\Delta H_s}{R} \qquad (6)$$

where

C = constant
m = moisture content
ΔH_s = latent heat of sorption
R = gas constant
m_1 = BET monolayer value

Equation (5), while useful in estimating m_1, is not appropriate for the coverage of the full range of water activities and many other equations have been proposed for this purpose. The most widely accepted is the GAB isotherm (Equation 7):

$$\frac{m}{m_1} = \frac{CKa}{(1 - Ka)(1 + (C - 1)Ka)} \quad (7)$$

where

K = constant
Other symbols have previously defined meanings.

As mentioned previously, the usefulness of (a) lies in its definition of conditions of equilibrium between phases, and its usefulness in defining the equilibrium moisture content of each component in multicomponent foods. The water activity of a multicomponent food is usually estimated assuming the validity of the Gibbs-Duhem equation, and this is most conveniently done by using the Ross equation (8):

$$a_m = a_1 \times a_2 \times \ldots . a_n \quad (8)$$

where

a_m = the water activity of the mixture
$a_{1..n}$= the water activity of each component 1 .....n if each were the only component interacting with all the water in the system.

Freezing Behaviour

Nucleation and crystal growth, which are the processes determining the ice distribution, will not be discussed here. Homogeneous nucleation is certainly an important property of water, but in food systems, heterogeneous nucleation is the most probable event, as an abundance of structures can act as centres for ice nucleation. Velocity of crystallization appears to be more dependent on the properties of the other components of the system - e.g. the properties of the macromolecular materials - than on the "properties of water" (Blond, 1985; Muhr & Blanshard, 1986).

The discussion will thus be focused on the "freezability" of water. This topic is clearly of direct importance in connection with several processes in food technology. The freezing behaviour of water reflects its general physico-chemical state in a material, and studies of freezing patterns add to our understanding of the mutual interactions between water and other food constituents.

State diagrams : The freezing behaviour of water in a solution may be illustrated by a state diagram, which represents the various states in which a system can exist as a function of temperature and concentration. Such diagrams result mainly from calorimetric measurements (differential thermal analysis = DTA, differential scanning calorimetry = DSC), but their elaboration may benefit from other experimental data including microscopic observation. State diagrams have been published for aqueous binary solutions of many biological components (Rasmussen & Luyet, 1969; Moreira, 1976). The general features of these diagrams are illustrated by the state diagram of glucose solutions given in Fig. 1. Since some variations in the determination of the transitions may be observed in the literature, examples of thermograms are also included.

The same type of diagram may be obtained with a solution containing several components. Fig. 2 represents the diagram obtained with blood plasma. It appears that the behaviour illustrated here should be representative of the freezing behaviour of a wide variety of food systems when no solute crystallizes. The curve M represents the temperature at which ice begins to separate when the system having the indicated concentration is cooled. This curve also represents the concentration of the remaining liquid phase as the temperature continues to decrease and more ice is separated from it. With some binary solutions, the freezing process ends when the residual liquid phase, having attained a specific concentration, fully crystallizes as a eutectic mixture of ice crystals and crystals of the solute. With most biological solutions, however, the solute does not crystallize during cooling; the concentrated liquid phase is solidified as a glassy material. The water remaining in this amorphous phase is described as unfreezable water.

Generally, the data which are used to construct the state diagram have been obtained during rewarming of samples previously frozen as fully as possible. This is because, very often, supercooling phenomena may occur during the cooling process. The first ice crystals then appear at a temperature lower than the temperature indicated by the M curve. More importantly, if freezing is rapid the remaining liquid phase is abnormally diluted. In that case, some ice is separated from the glass phase during rewarming, giving rise to a devitrification exotherm (D) on the thermogram. The amplitude of this phenomenon depends on the cooling rate : a rapid cooling does not give enough time for water molecules to diffuse towards the crystallization interface.

The presence of the glass phase in the system is revealed on the thermograms by a change in specific heat (G). Several other changes in physical properties of the material are associated with a glass transition, namely a change in mechanical properties. It is generally admitted that the glass transition corresponds to a border between the solid and liquid states, i.e. to an isoviscosity line of 10^{14} N s m^{-2}.

Whatever the initial concentration of the system, melting of ice always begins at the same temperature (temperature of incipient melting, T_{IM}). The temperature of the glass transition, which may take a lower value (T_G) when the glass phase is over diluted, as indicated by a devitrification exotherm, is otherwise equal to a constant value (T_{G1}). It can thus be concluded that, either by slow cooling, or by devitrification during rewarming, a glass phase of constant composition may be formed, corresponding to a defined content of unfreezable water. Samples with a lower water content do not exhibit ice formation during cooling or rewarming.

In most biological systems, the glass phase appears quite stable, the contained water not being observed to separate as ice even during prolonged storage at low temperatures or during repeated annealing treatments. It must be stressed, however, that this stability may be a relative concept. When the solute is able to crystallize (as is the case with sugars) it is not impossible that finally, after repeated annealing processes for instance, a eutectic mixture is formed (Young & Jones, 1949).

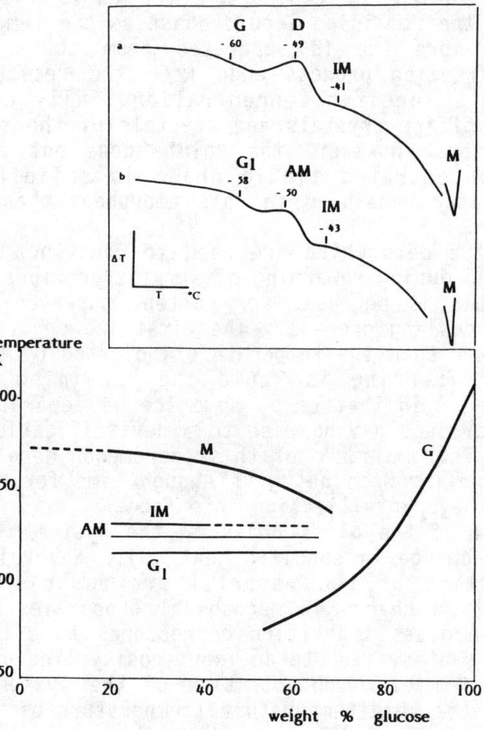

Figure 1. State diagram of glucose-water (from Moreira, 1976). Insert: Thermograms (DTA) of a glucose solution (concentration 40%); a - direct rewarming, b - after a first rewarming limited to -49°C.

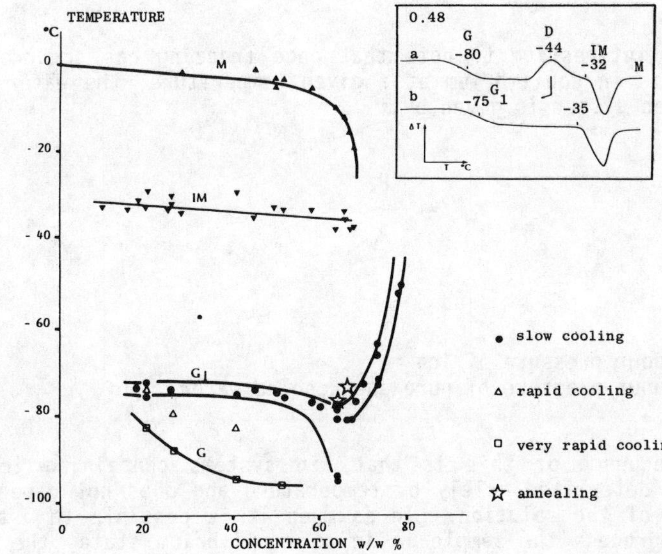

Figure 2. State diagram of blood plasma (from Simatos _et_ _al_., 1975). Insert : Thermograms (DTA) of plasma (water content = 0.48 g/gds); a - direct rewarming, b - after annealing at -44°C.

Freezing _behaviour_ _and_ _water_ _activity_ : The depression of the freezing point - which is represented by the M curve on the state diagram - is a colligative property and, for an ideal solution, is given by the following expression (heat capacity changes being neglected) :

$$\ln X_W = \frac{L}{R} \frac{T_O - T_M}{T_O T_M}$$

where

X_W = molar fraction of water in the solution
T_O, T_M = freezing temperatures of water and solution
L = latent heat of fusion at the freezing point of water (T_O)

For a non-ideal solution, a similar relationship exists between the freezing point and the activity of water :

$$\ln a_w = \frac{L}{R} \frac{T_o - T_M}{T_o T_M}$$

It is interesting to note that once freezing has occurred and if the sample is in equilibrium at a given temperature, the water activity of any frozen system is given by :

$$a_w = \frac{P_i}{P_w}$$

where

P_i = vapour pressure of ice
P_w = vapour pressure of pure supercooled water

and a consequence of this is that, in systems containing ice, water activity is determined solely by temperature and does not depend on the composition of the solution. In as much it is possible to assume that during the process the sample is in an equilibrium state, the DSC scan between temperatures T_{IM} and T_M represents the quantity of ice melting as a function of temperature, that is as a function of water activity. It should be noticed that the relationship between a_w and $(T_o - T_M)$ is valid whatever is the origin of the depression of water activity in the system: presence of solutes, and the various types of interactions between solutes and water affecting a_w, but also structural effects.

According to the Kelvin equation, $\ln a_w$ is inversely proportional to the radius of a capillary in which the water is contained. It has been demonstrated (Brun et al., 1979) that the DSC curve obtained with water or benzene in a porous material (alumine - Millipore filter) can be analysed so as to provide a pore distribution spectrum (repartition of porous volume as a function of pore radius) which is in satisfactory agreement with the spectrum which can be obtained with other methods of analysis. One may thus be tempted to use DSC curves to determine pore size in food systems such as gels.

Various experimental observations suggest that several fractions may be distinguished in freezable water. In the case of caseins, Ruegg et al. (1974) claimed that a small fraction may be recognized within the range of water content immediately above the limit of unfreezable water, differing from the bulk of the freezable water in exhibiting a latent heat of fusion different from that of pure water. Most often, the melting endotherm in DSC scans appears to comprise two or several more or less resolved peaks. Ikada et al. (1980) attributed such peaks observed with mucopolysaccharides to fractions of water characterized by weaker interactions (as compared to unfreezable water) with hydroxyl and ionic groups.

Nature of unfreezable water : Consideration of the state diagram (see above), as well as of the variation of the latent heat of melting as a function of water content, leads most authors to conclude that the unfreezable water content of a system is a fixed quantity independent of total water content. However, several reports in recent years (Biswas *et al*., 1975; Ross, 1978; Carlsson *et al*., 1986) have claimed that the unfreezable fraction of water increases in size as the system becomes more fully hydrated. In the case of polymeric materials, it is possible that such increases can be satisfactorily explained in terms of conformational changes in the polymer (Carlsson *et al*., 1986). It is probable, however, that most often the observations could be explained by incomplete crystallization of freezable water or by the failure to take account of the contribution of the heat of dilution as ice melts into the concentrated solution.

A remarkable agreement has been demonstrated with solutions of various sugars between the unfreezable water content and the quantity of water immobilized through interactions with the solute, as determined by NMR or dielectric relaxation measurements (Franks, 1983). However, it is emphasized that "the residence time of a given water molecule at the solvation site is short and the water must therefore not be considered as being bound". The view is now widely accepted that the energy of interaction between water and any solute cannot be stronger than the hydrogen bond between two water molecules in ice (Franks, 1985b). The state of unfreezable water thus appears to be determined by kinetic factors more than by energetic ones and it is claimed (Franks, 1985b) that the unfreezable water content should correspond to the intersection of the M curve and of the glass transition curve (which corresponds to the temperature at which the viscosity reaches 10^{14} N s m^{-2}). Although this explanation presents some convincing aspects, several features of the state diagrams remain to be discussed.

Capability to Act as a Solvent

Dissolution of a solute may be simply defined as the process of its dispersion at the molecular level. Mobility is associated with this dispersed state when the system is a liquid.

For a solute to be dispersed in the solvent, the transfer of a solute molecule from the pure solute to the solvent must correspond to a favorable change in free energy ($\Delta G < 0$). This negative ΔG is the algebraic sum of the enthalpic and entropic terms which describe the various changes in structures or in intermolecular bonds accompanying the solute transfer; these changes include : interactions between solute and water which replace solute-solute interactions, changes in water structure, and eventual changes in the solute configuration.

Of course, the solutes which may be the more easily dispersed by water are solutes whose chemical structures comprise a high proportion of polar or ionic groupings. It is worthwhile to remember however that, besides the number of these groups, the molecular configuration is important (e.g. sugars). Also, substitution of some polar groups by apolar ones may make the material more soluble in water : in cellulose the regular configuration and the numerous OH groups result in a crystalline insoluble material, whereas substitution of some OH by methyl groups permits its solubilisation.

It is often stated that not all the water in a moist biological system exhibits "normal" solvent properties. Actually, the concept of "non-solvent water" has been used in two very different experimental contexts which correspond probably to different properties of water.

The classical method, which has been very widely used to estimate the non-solvent water content in many biological systems, was based on the addition of a test solute to a **system already rich in water**; then the concentration in the liquid phase was determined either directly after extraction of a sample of this liquid phase if possible, or by measuring the change in some colligative property, usually freezing point depression. From this concentration, the quantity of water present in the solution phase was calculated and compared to the total water content, assuming that there were two distinct water fractions in the system, one which was not free to act as a solvent, the other one in which the test solute was assumed to be uniformly distributed (see Duckworth, 1981 for a review).

In the 1970's a non-solvent water content was defined as the water content level **in a dehydrated product** above which the mobility of a test solute becomes higher than a given level. Two techniques were used : nuclear magnetic resonance (NMR) and electron spin resonance (ESR).

The NMR work (Duckworth & Kelly, 1973; Duckworth et al., 1976; Duckworth, 1981) employed a simple wide-line instrument capable only of distinguishing "liquid state" hydrogen protons. The signal from the former, which included contributions both from water and from hydrogen-containing compounds dissolved in it, was measured, but without the possibility of separating the contributions from the two sources. Numerous individual investigations, performed on systems composed of a polymer and a small molecular weight solute (as well as with more complex food systems), showed a common pattern of behaviour : with increasing moisture level starting from the dry polymer, a range of water content was first encountered over which the addition of the small hydrogen-containing solute produced no incremental increase in signal size over that due to the "liquid state" water. This was followed by a range, beginning at what was termed the "mobilisation point" of the test solute, over which an increase in signal size attributable to the solute was measurable, the increment itself increasing to a maximum at some characteristic higher moisture content, after which the increase due to the solute was constant and in close agreement with the theoretical increase corresponding to complete solution of the quantity of solute introduced (Fig. 3).

ESR (Simatos et al., 1981; Voilley & Le Meste, 1985; Le Meste & Duckworth, 1987) requires the use of paramagnetic probes, e.g. stable nitroxide free radicals. Such probes may be chosen so as to be more or less similar in size and chemical properties to common food constituents, but the presence of the nitroxide radical is necessarily peculiar to themselves. The probes being added to polymer systems, the ESR studies, similarly to the NMR experiments, have made it possible to define a moisture level above which a specific probe begins to be mobilized. Moreover, the ESR spectra demonstrated that the proportion of mobile probes increased progressively within a given range of the water content above the mobilisation point (Fig. 4).

Although direct experimental evidence is scarce, it is probable that the role of water in the experiments with the dehydrated materials is a plasticizing effect. The water content identified with the mobilisation point should be the water content at which the temperature of the glass

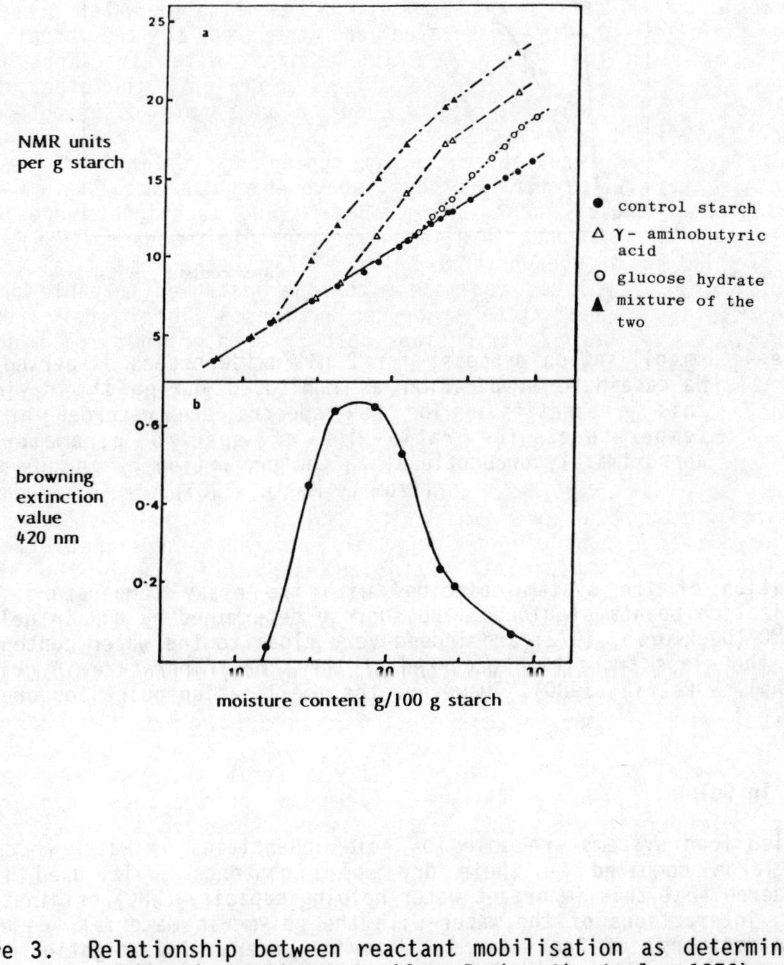

Figure 3. Relationship between reactant mobilisation as determined by NMR and reaction rate (from Duckworth <u>et al</u>., 1976).

a - Mobilisation of γ -aminobutyric acid, glucose and a mixture of the two in hydrated gelatinised starch.

b - Browning extinction values of water extracts of the samples used for the readings in (a) after 7 days at 5^{0}C.

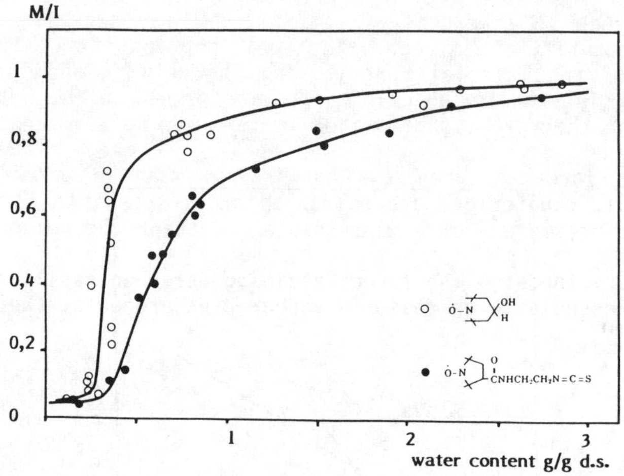

Figure 4. Immobilisation process for 2 nitroxide probes dispersed in a Na-caseinate preparation as monitored during the drying of this preparation. The ESR spectra are recorded at room temperature. The ratio M/I of spectra parameters is approximately proportional to the proportion of mobile probes (from Le Meste, 1986).

transition of the system coincides with the essay temperature. The mobilization points of glucose and sucrose determined by NMR in gelatine at 23°C (Duckworth, 1972) are indeed very close to the water content for which the glass transition occurred at the same temperature in gelatine (Marshall & Petrie, 1980). However, the mobilisation point for urea was lower.

Water in Gels

Gelified food systems are able to "hold" quantities of water which are very large compared to their dry solid contents. It used to be considered that this important water holding capacity (WHC) originates in strong interactions of the water with the polymeric material, resulting in "structuring" of the water. However, several investigations using dielectric relaxation time (De Loor & Meijboom, 1966) or N.M.R. (Suggett, 1975) have demonstrated that a very large proportion of the water retained in gels has the same rotational mobility as pure water. Only a small quantity of the water in the gel (e.g. 0.5 g water/g agar - Suggett, 1975) has a restricted mobility and may thus be considered as "bound". The unfreezable water content is not significantly different in gelified and non-gelified homologous systems (Blond, 1987). It is possible, however, that some "binding" of water results specifically from the gel structure, by trapping between aggregated chains, as has been shown with gelatine (Maquet et al., 1986).

The most important part of the water is thus held by loose forces. In the presence of an excess of water, a gel may swell by absorbing an additional quantity of solvent. The swelling capacity is limited by the elasticity of the material constituting the network and the number and strength of interchain junctions. The forces which induce the gel swelling - or the retention of water in it - may be assumed to be:

- Osmotic forces. When the network is made of a polyelectrolyte material, counterions are entrapped and responsible for an elevated osmotic pressure - or a depressed a_w - inside the network.

- Capillary forces. The force retaining water in capillary spaces may be expressed as a decrease in vapour pressure or a_w (Kelvin law):

$$\ln a_w = \frac{V_w}{RT} \gamma \cos\theta \left(\frac{1}{r_1} + \frac{1}{r_2}\right) \quad (1)$$

or as the pressure necessary to counteract the capillary rise, h :

$$h = \frac{\gamma}{\rho g} \cos\theta \left(\frac{1}{r_1} + \frac{1}{r_2}\right) \quad (2)$$

where

γ = surface tension of the liquid
V_w = molar volume of water
ρ = density of the water
R = gas constant
T = temperature (K)
θ = wetting angle
r_1, r_2 = radii of the capillary and of the meniscus

According to Griffin (1981), the water potentials resulting from osmotic and capillarity effects may be taken to be simply additive. The total water potential, Ψ , defined by:

$$\Psi = \frac{\mu_w - \mu_w^o}{V_w} = \frac{RT \ln a_w}{V_w} \quad (3)$$

is thus equal to

$$\Psi = \Psi_s + \Psi_m \quad (4)$$

where Ψ_s corresponds to the solute (or osmotic) effects and Ψ_m to the matrix (or capillary) effects.

$$\Psi_m = \gamma \cos\theta\left(\frac{1}{r_1} + \frac{1}{r_2}\right) \qquad (4)$$

The relationship between the WHC and the microscopic structure of gels has been very well documented, on a qualitative basis (Hermansson & Lucisano, 1982; Hermansson, 1986). It has been demonstrated with blood plasma protein and whey protein gels submitted to various treatments (temperature, pH, etc.) that the finer the structure, as observed by scanning electron microscopy, the higher the WHC.

It would be very interesting to relate quantitatively the microscopic structure and the WHC of gels, and to check experimentally the applicability of Equation (4) to food gels. Some authors have doubted that the Kelvin law remains valid at very small values of r_1 and r_2. An experimental verification has been presented for cyclohexane condensed onto mica surfaces with radii r_2 as low as 4 nm (Fisher & Israelachvili, 1979). It may be further objected that this verification has been obtained with a very rigid system and one may still wonder whether Equation (4) can be applied to food gels which are highly deformable. There may also be important differences between heterogeneous coarse gels which have rigid, rather permanent structures (e.g. agar gels or highly matured gelatin gels) and homogeneous transparent gels with much finer structures where the existence of pores with permanent size or location is difficult to imagine.

Moreover, adequate values for the different terms of Equation (4) are uncertain. Most often, the wetting angle θ is assumed to be = 0 (and then $r_1 = r_2 = r$). But the value of r is very difficult to determine. Several methods of two main kinds probably need to be used in order to arrive at reliable results:

1. Microscopic methods (scanning electron microscopy and transmission electron microscopy) (Hermansson & Buchheim, 1981) which provide quantifiable images, but which involve drastic procedures for the preparation of specimens;
2. Methods which are less prone to induce alterations of the sample structure, but which give only indirect information on the pore size distribution, such as diffusion of macromolecules (Busk & Labuza, 1979), NMR (Lillford <u>et al</u>., 1980) or laser light scattering (Tanaka, 1981).

EFFECTS OF WATER ON PROPERTIES OF FOOD COMPONENTS

Structuring Functions of Water

The ability of water to hydrogen-bond with itself as well as with other molecules has a significant impact on the structure and conformation of other food consitutents. Major food components which are so affected are proteins, polysaccharides and phospholipids.

A recent review by Rupley et al. (1983) summarizes well the effect of water on the structure and function of globular proteins. These authors emphasized the stepwise nature of protein hydration, which has been recognized by most previous investigators. They divided the hydration process into 4 steps. In the case of lysozyme, the first step corresponds to 0 - 0.07 g H_2O/g protein, and involves mainly binding of water to charged groups. This water imparts very little mobility to the protein and enzyme activity is negligible. The second step corresponds to 0.07 to 0.25 g H_2O/g protein, and completes hydration of polar groups leading to some clusters of water. The protein mobility increases by a factor of 1,000 over this range which merges into the next two steps in which full solution properties of the protein are attained.

A possible allosteric significance of water in protein has been postulated by Mackay & Wilson (1986), who studied molecular dynamics of ion solvation and transport through the ionophore gramicidin A. They observed that water forming a linear single file within the channel has different solvation and dynamic properties as compared to bulk water. They proposed that protein motion may act as a "switch" on the water in the transport channel and this water, in turn, acts as a "hydraulic plunger". They concluded that "water networks exist inside proteins, sometimes at critical locations" and believe "that a dynamic role for these water structures in enzyme function should be considered".

Of interest in relation to protein-water interactions is also the recent work of Wolfenden et al. (1981) who measured affinities of different amino acid side chains for solvent water. They reported their results as equilibrium constants for transfer of amino acid side chains (RH) from water (at pH 7) to vapour :

$$(k_{eq}) = \frac{\text{RH(vapour)}}{\text{RH(aqueous)}}$$

They found k_{eq} ranging from 10^2 for the nonpolar side chains (glycine, alanine) to 10^{-8} for most ionizable side chains to 10^{-14} for the extremely hydrophilic arginine.

The idea of water and protein segments interacting to form a unique, water-containing structure is not new and has been proposed for collagen by Berendsen in 1975. Effects of water on folding of protein were reviewed by Eisenberg & MacLachlan (1986).

Similar water-polar group interactions are known to occur in polysaccharides and, in particular, in starch and gums (Maurice et al., 1985). In most cases, the role of water is that of plasticizer and is reviewed elsewhere in this paper.

Finally, it should be noted that water in the form of layers, either mono- or multimolecular, is an important component in structuring polar lipids and, in particular, phospholipids. This subject is too complex to review here and the reader is referred to the recent authoritative work of Small (1986).

Diffusional Properties of Small Molecules

Fundamentals : Translational diffusion of solutes under the influence of a concentration gradient, as described by Ficks's law, is an important phenomenon which, besides controlling the rate of various extraction operations, governs several processes occurring during drying and storage of low moisture foods such as redistribution of solutes, loss of volatile aroma compounds, and chemical and biochemical reactions. Rotational diffusion does not seem to have such direct practical implications. However, it is an experimental parameter which has proven very useful because it is closely related to translational diffusion and may be used to estimate the mobility of solutes.

Numerous methods have been described for the measurement of translational diffusion coefficients, which are based on solutions of Ficks's law, which can be expressed by :

$$d_{mB} = - AD_{CB} \frac{\partial_{CB}}{\partial x} dt$$

where

d_{mB} is the quantity of solute diffusing across the small surface A during the time dt.

$\frac{\partial_{CB}}{\partial x}$ is the concentration gradient in the considered direction.

D_{CB} is the translation diffusion coefficient of B at the concentration C_B .

For measurement of translation diffusion coefficients in liquids of low viscosity, satisfactory methods are available (e.g. Stokes' cell). For measurements in concentrated, highly viscous solutions, special methods have been developed (capillary tubes, entrapment of the solution in a gel and determination of a concentration profile) which remain tedious and may present practical problems. In solid systems with low moisture contents, measurements are very difficult because of the low value of D, although the use of radiographic tracers as solutes has proved capable of providing very useful data (Duckworth, 1962; Duckworth & Smith, 1963).

Rotational diffusion constants (D_{rot}) represent the frequency of reorientation of molecules which can be deduced from various techniques : dielectric relaxation, fluorescence depolarisation, electron spin resonance (ESR), etc. D_{rot} is related to the correlation time (τ_c) which may be defined as the length of time over which the molecules persist in a given orientation.

$$D_{rot} = \frac{1}{6\tau_c}$$

The correlation time can be determined very easily by ESR (or at least evaluated when the solute mobility is very low). The drawback of this method is that the measurement is made on paramagntic probes which have to be dissolved in the sample and that these probes necessarily differ in detailed molecular structure from normal food components.

It may be noticed that in heterogeneous systems (particulate or porous systems, or those with a cellular structure), the measured D_{trans} is an apparent diffusion coefficient which reflects the influence of the structure. On the contrary, D_{rot} may be more representative of the local conditions.

The translational and rotational diffusion coefficients for a particle in a liquid are related to the frictional coefficient ξ by the Einstein equation:

$$D = \frac{kT}{\xi}$$

where

k = Boltzamann constant
T = temperature (K)

According to the hydrodynamic model for a spherical particle of radius, r:

$$\xi_{trans} = 6\pi r\eta$$

$$\xi_{rot} = 8\pi r^3\eta$$

where η is the viscosity of the solvent.

The following equations can thus be derived:

$$D_{trans} = \frac{kT}{6\pi r\eta} \qquad \text{(Stokes - Einstein equation) (1)}$$

$$D_{rot} \equiv \frac{kT}{8\pi r^3\eta} \qquad (2)$$

Experimental results are described by these equations more satisfactorily when the diffusing solute is a large molecule than when it is a small one. They are, however, currently used, although in the latter case the equation proposed by Wilke & Chang (1955) for translational diffusion should be more adequate :

$$D_{trans} = 1.17\ 10^{-16} \frac{(\varphi M)^{0.5}}{V^{0.6}} \frac{T}{\eta} \quad (3)$$

where

V = molar volume of the solute
M = molecular weight of the solvent
φ = association parameter of the solvent (2.6 for water)

This equation applies for dilute solutions of non-electrolytes when solvent and solute have similar molecular weights (Loncin, 1980).

Similarly, for the rotational diffusion coefficient, r should be interpreted as representing the effective hydrodynamic radius of the diffusing molecule, taking account of its interactions with molecules of the solvent. When the rotating molecule has a size comparable to that of the solvent molecules, it is suggested (McClung & Kivelson, 1968; Dote et al., 1981) that the value of r depends on the prevailing boundary conditions between two extremes : stick conditions if the solvent molecules stick to the surface of the rotating molecule (as is assumed in the hydrodynamic model), and slip conditions if the rotating solute displaces the surrounding fluid which exerts no drag force on the particle.

Diffusion in food as a function of water content : At low water contents, diffusion is severely limited. In a pioneering study, Duckworth & Smith (1963) measured diffusion of radiolabelled glucose and sulfate in pieces of dry vegetables. They found that the lowest level of moisture at which diffusion was detected was about 1.3 times the monolayer value. Considerable work has been done in connection with diffusion of water and of flavours during drying. Fig. 5 illustrates the effect of moisture content on the diffusion coefficient for water and shows clearly the drop in this coefficient at low moisture contents. In addition, the activation energy for diffusion shows a rapid rise at low water content (Fig. 6). The diffusion coefficient for organic compounds drops even more rapidly than that for water as shown in Fig. 7.

The rotational correlation time of nitroxide probes, which is in the range of 10^{-10} - 10^{-9} s in casein systems with 0.5 to 3 g water/g d.w., becomes larger than 10^{-7} s for water contents below 0.2 g/g. In the latter case, the probes have been characterised as "immobilised" because the ESR spectra are representative of "powder spectra". The probes then have only slow tumbling motions, the rate of which decreases only slowly with water content (Simatos et al., 1981).

In concentrated solutions, as can be observed in Fig. 5, the variation of the diffusion coefficient is rather small when measured in different food systems, at a given moisture content. This fact was also reported for the diffusion of sorbic acid in very different food systems (Giannakopoulos & Guilbert, 1986) and for the diffusion of several volatiles in solutions of various sugars and maltodextrins (Voilley & Roques, 1987).

The variations of diffusivity as a function of water content, or for a given water content with the nature of the "substrate", originate mainly from the variations in viscosity. This effect of viscosity is

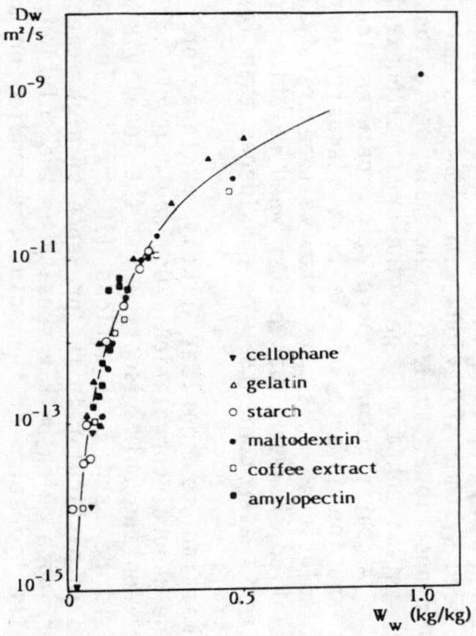

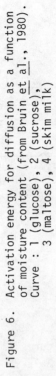

Figure 5. Water-diffusion coefficients in some food materials as a function of moisture content (from Bruin et al., 1980).

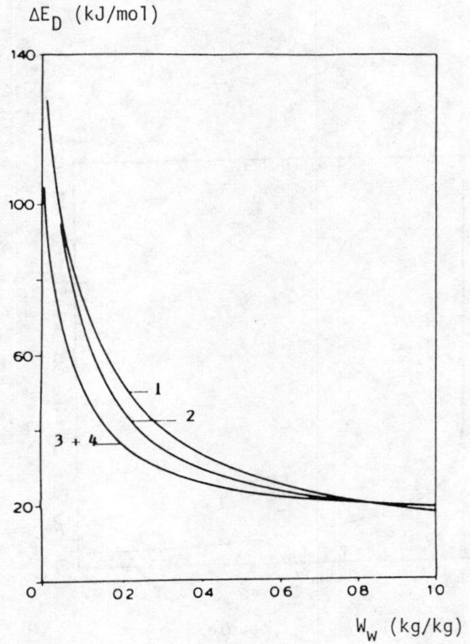

Figure 6. Activation energy for diffusion as a function of moisture content (from Bruin et al., 1980). Curve : 1 (glucose), 2 (sucrose), 3 (maltose), 4 (skim milk)

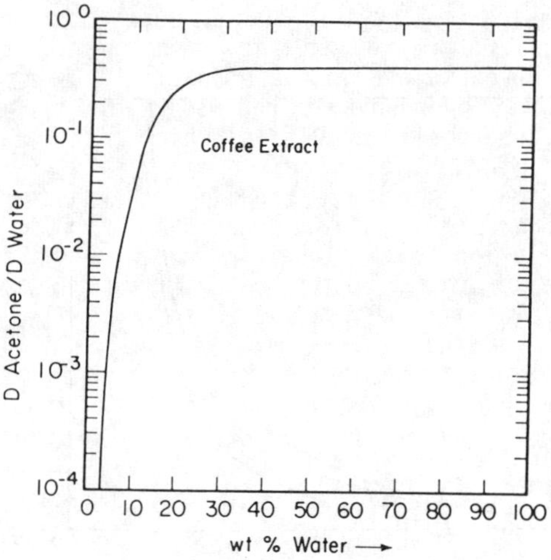

Figure 7. Ratio of diffusion coefficients of acetone and water in coffee extract as a function of water content (from Thijssen & Rulkens, 1968).

predicted by the Stokes-Einstein equations as well as by that of Wilke & Chang (1955). The relevant viscosity, however, should be the true viscosity prevailing in the near environment of the diffusing species.

Guilbert et al. (1985) reported that the D_{trans} measured for sorbic acid in different food systems of various water contents and compositions was approximately proportional to $1/\eta$, the main deviations from this correlation being systems containing gelatine or a maltodextrine of high molecular weight. A satisfactory correlation between the D_{trans} of several volatiles and the inverse of the bulk viscosity has also been demonstrated for solutions of sugars and maltodextrins (Bettenfeld, 1985). The correlation, however, was better when solutions of the same solute with different water contents were considered than for solutions of various solutes.

It is interesting to consider likely explanations for these departures from the general correlation. One could invoke :

- discrepancy between the bulk viscosity and the local viscosity - the one which should be used in Equations (1 - 2). This explanation should be valid in particular in the presence of high molecular weight solutes, conferring a high bulk viscosity on the solution;
- variations in the hydration of the solute, depending on availability of water in the solution as represented, for instance, by water activity;
- interactions between the diffusing species and other components of the system.

Parallel measurements of D_{rot} and D_{trans} on a nitroxide probe may help in analysing these effects, as carried out by Le Meste & Voilley (1987) in solutions of sugars and maltodextrins with a fixed water content (1 g/g d.w.) :

- the relatively high value of the product $D_{rot} \cdot \eta$ observed in the glucose solution as compared to the value in the maltose solution, which could not be explained in these solutions by a viscosity effect, suggested a smaller hydrodnyamic radius of the probe in the glucose solution (shift toward slip condition) which correlated with the lower water activity of the solution;
- the ratio D_{trans}/D_{rot} should not depend on the viscosity or the nature of the solvent, but only on the probe radius. Its higher value in maltodextrins solutions with higher molecular weights suggested again a greater degree of hydration of the probe in these solutions (stick conditions) correlating with a higher water activity.

It may be stressed that in solid systems, the influence of water on mobility of solutes should result from a different mechanism from the afore mentioned one for solutions. In this case, the effect of water should be to rupture bondings between macromolecular components of the system and to impart flexibility to them, thus permitting the mobility of small solutes.

Dynamic Properties of Macromolecules

Although displacements of a molecule as a whole (translation and rotational diffusion) could be envisaged under this title, this section will deal only with internal motions of macromolecules. Full realisation of the importance of molecular mobility in relation to the properties and functions of macromolecules, in particular proteins, has only become widespread in recent years. This mobility, in terms of its influence on enzyme action and other "functional" properties, is extremely important in relation to several aspects of food quality and stability which are discussed in other sections. Here we are concerned with the immediate effects of water on the dynamic properties underlying the above-mentionned events.

From the abundant literature on the subject (e.g. Gurd & Rothgeb, 1979; Karplus, 1986) internal motions in a globular protein may be presented in a simplified view as :

- Thermal displacements of atoms and reorientations of amino-acid residues, the mobility of the nonpolar aromatic residues being lower than that of the polar groups, frequency and amplitude of the motion increasing from the centre to the periphery of the molecule.
- Relative displacements of protein domains or subunits.
- Microopening of the structure.

Amplitudes of motion may range from 0.01 to 100 Å, energies from 0.1 to 100 kcal, and times from 10^{-15} to 10^{3} s.

The influence of water on these molecular motions does not seem to have been extensively studied. From a study by molecular dynamic simulation carried out on pancreatic trypsin inhibitor, Van Gunsteren & Karplus (1981) concluded that the primary effect of the water was to stabilize an average structure closer to the native one. There was also a dynamic effect, i.e. a damping of the positional fluctuations of surface and some interior atoms. Glass transition has been identified in globular proteins by calorimetric experiments, measurement of thermal expansion coefficient, or measurement of viscoelastic properties (Young's modulus)(Morozov & Gevorkian, 1985; Stein, 1986). The proteins, in particular lysozyme, bovine serum albumin and myoglobin, were studied in the solid state as crystals or amorphous films. A glass transition was observed in the temperature range 150 - 220 K for samples equilibrated in relative humidities ranging from 0 to 97%. The transition in the wet samples was associated with mobility of the surface protein groups and "bound water molecules", whereas in the dehydrated material, the relaxation process was assumed to be localized in the protein interior.

In an attempt to study the limiting effect of lower levels of hydration on molecular mobility, an investigation has been carried out on casein using ESR (Le Meste, 1986). Two spin probes were attached to sodium caseinate molecules via the reactive groups of amino acid side chains (alcohol, amine and thiol), the major part of the linkage involving the NH_2 groups of lysine residues. The evolution of mobility of these probes both when covalently linked to the protein ("spin labels") and when free in solution ("spin-probes" s. str.) was studied in sodium caseinate concentrates over a wide range of water content.

In general, the measured mobility of the spin labels was lower than that of the probes. Whereas the free probes were completely mobilised in the case of probe II at water contents above 0.7 g water/g casein, immobile radicals were still present among covalently linked labels at water contents as high as 3 g/g (Fig. 8). The correlation time for the mobile radicals was higher at all water contents for the labels than for the free probes.

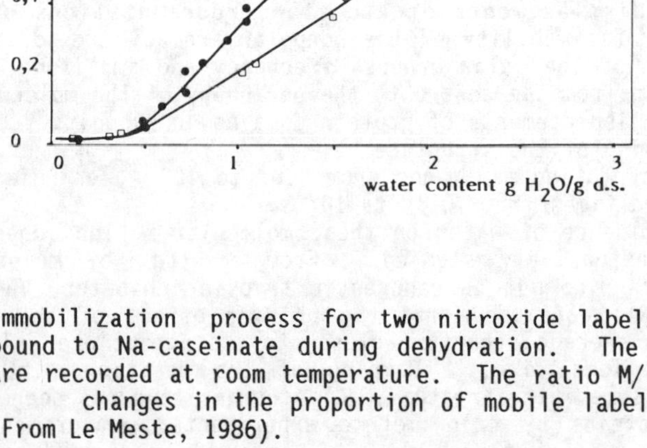

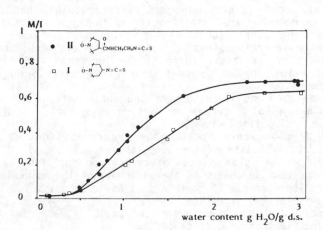

Figure 8. Immobilization process for two nitroxide labels covalently bound to Na-caseinate during dehydration. The ESR spectra are recorded at room temperature. The ratio M/I represents changes in the proportion of mobile labels. (From Le Meste, 1986).

Comparison of the behaviour of the two different radicals in the covalently-linked state indicated that radical I, although of a smaller size, had a reduced mobility. This probably reflected the fact that, because of its smaller size, this radical was able to react with residues more deeply buried in the protein structure.

CONSEQUENCES ON SPECIFIC PHENOMENA RELATED TO FOOD QUALITY

Rheological Properties

Reviews have been recently devoted to the flow properties of solutions and dispersions of polysaccharides (Launay et al., 1986) and of proteins (Rha & Pradipasena, 1986). Here will thus be presented very briefly the points which concern particularly the hydration of these solutes. Information relevant to solute-solvent interactions in dilute solutions may be obtained from the value of the exponent α in the Mark-Houwink equation :

$$[\eta] = KM^{\alpha} \qquad (1)$$

where $[\eta]$ is the intrinsic viscosity of the solution and M is the molecular weight of the solute.

Different ranges of α values may be predicted according to the hydrodynamic behavior of the macromolecule (Flory, 1953; Mitchell, 1979):

- In the equivalent sphere model, the solute molecules are regarded as particles including a given volume of solvent which is associated hydrodynamically with each of them. The volume fraction of the particles Φ and the intrinsic viscosity are given by the following expressions :

$$\Phi = c(v_2 + h_1 v_1) \qquad (2)$$

$$[\eta] = \beta(v_2 + h_1 v_1) \qquad (3)$$

  where

  v_2, v_1 = specific volumes of the solute and solvent
  h_1 = weight of solvent associated per unit weight of solute
  c = concentration
  β = shape factor of the particles (= 2.5 for spherical, rigid, non-charged and non-interacting particles).

The term $(v_2 + h_1 v_1)$ is sometimes named the voluminosity of the solute. For globular proteins, the volume of solvent which must be regarded as associated with the molecule is independent of the molecular weight. It follows that the intrinsic viscosity is independent of molecular weight ($\alpha = 0$). For flexible chain molecules for which the equivalent sphere model is relevant, α may be demonstrated to be comprised between 0.5 and 0.8. It should be close to 0.5 in poor solvents and approach 0.8 for very good solvents. For rod-shaped macromolecules, if assumed to be impenetrable to solvents, α should be close to 1.8.

- In the free draining molecule model, the solvent flow behaviour is assumed to be unperturbed by the presence of the polymer. In this case, it is possible to show that α should be = 1 or slightly higher than 1 when the solute is a spherical molecule and = 2 for a rod-shaped polymer.

"Hydrodynamic hydration" values have been calculated using Equation (3) (Kuntz & Kauzmann, 1974). These authors have emphasized that the reported values are uncertain, because of difficulties in obtaining adequate values for v_2 and for β, in the case of non-spherical molecules. The dehydration for different proteins was estimated to be between 0.2 and 0.5 g water/g protein, which is about the same range (or a little higher) as the range for the unfreezable water content.

An interesting point is that, for denatured proteins, $[\eta]$ is much higher than for the native globular proteins (corresponding to values of 0.5-0.8 instead of 0). On the countrary, denaturation is accompanied by only a very limited increase in the unfreezable water contents. The same observation can be made in other circumstances where the sphere equivalent to a flexible coil molecule is expanded. The intrinsic viscosity of casein in which some lysine NH_2 residues have been substituted with galactose has been demonstrated to be higher than in original casein and to increase with the substitution rate (Colas, 1987). The unfreezable water content was not significantly changed (Blond, 1987).

Plasticization, Glass Transition and Stability

Water is the important plasticizer for food components. As a consequence, the mobility of polymer chains and the ease of diffusion of small molecules are affected by water content. The most important properties of foods or technological processes which are affected by this aspect of hydration include :

- Texture of dry or intermediate moisture foods.
- Diffusion in and through foods of various molecules. This aspect is of enormous importance in such operations as salting, osmotic treatments, transfer of flavours between food components and, of course, in all the processes for food dehydration.
- Reactivity of groups on food polymers, in particular enzyme activity.
- Phase transitions of particular importance such as those affecting starch and the amorphous glasses formed by sugars and sugar-polymer mixtures.

In the present section, only the phenomena which affect the physical properties of the product will be reviewed, chemical and enzymatic reactivity being discussed in the following section.

Texture : The literature is replete with experimental results relating rheological properties such as maximum force before failure in compression, or "hardness", or the elastic modulus, to water content. Studies which specifically relate changes in texture to significant, and universally valid levels of water content, or water activity, are few. The following generalizations seem to be valid, however.

In the absence of actual phase transition affecting the arrangement of structural elements in food, the main function of water is plasticization of polymer chains (proteins, starch, other polysaccharides). It would be expected that hydration below the monolayer level would produce no textural changes and that above this level the textural properties would indicate a "loosening of structure" and that these changes would increase with increasing water content. Probably the best study demonstrating that these expectations are correct is that of Katz & Labuza (1981). In this study, various snacks including saltine crackers, potato chips and popcorn were evaluated. BET monolayer values and rheological and sensory measurements were also obtained. Table 1 is based on the results of these authors. They concluded that the significant textural changes occur well above the monolayer value. While we agree with this conclusion with respect to water activity, it should be noted that all the transitions occur in the portion of the isotherm which has low ($\Delta m/\Delta a$) slope. Thus it seems that the critical values that were obtained here correspond to levels of absorbed water of between 1.5 and 2 times the "monolayer". It seems very reasonable to assume that plasticization of polymer chains would require this much water above that bound to primary sites. Katz & Labuza (1981) referred to the possibility of transition in the amorphous sugar in the snacks and this, in fact, may be a factor which in itself is related to saturation of primary sites for adsorption, followed by provision of additional, "mobilizing" water molecules.

TABLE 1
Rheological and sensory critical values of water content and water activity of various snacks (based on Katz & Labuza, 1981)

Parameter	Critical Ranges of Water	
	Water Activity	Water Content
	Saltine Crackers	
Initial Slope	0.1 - 0.35	0.03 - 0.06
Peak force	0.52	0.08
Hedonic Crispness	0.39	0.07
BET Monolayer	0.22	0.05
	Popcorn	
Work for 75% Compression	0.20	0.04
Hedonic Crispness	0.49	0.06
BET Monolayer	0.17	0.03

Diffusion : The fact that the diffusion coefficient for organic compounds drops rapidly at low moisture contents, even more rapidly than that for water, is the basis of the "selective diffusion" theory which was formulated by Thijssen (1979). This theory, with some modifications, is generally accepted as the basis for encapsulation and flavour retention in drying.

Phase transitions : Most work in this area has been centred on spray-dried and freeze-dried materials which produce amorphous sugar glasses entrapping in them various volatiles, and on transitions in starch. We shall illustrate the state-of-the-art by discussing freeze-dried materials.

The transitions in freeze-dried materials have been studied in particular by the groups of King et al. (1976), Thijssen (Rulkens & Thijssen, 1972), Karel (Tsouroflis et al., 1976), and Simatos (Le Meste et al., 1979), and are usually described as "structure collapse". Most recently, this subject has been reviewed by Levine & Slade (1986). In freeze-drying, water is removed by sublimation of ice. If the initial solids concentration of the starting solution is greater than 1% (weight basis), the freeze-dried solid will retain the shape of the container in which it is frozen, thus forming a cake. Other methods of drying produce powders, as in spray-drying, or sheets, as in drum-drying. Only in freeze-drying is it possible to obtain an extensive gross structure of low bulk density. Microscopic examination of freeze-dried substances shows layers of needlelike void spaces previously occupied by the ice crystals that formed during freezing.

When the freeze-dried cake is subjected to heating, at a certain temperature a change in structure known as collapse generally occurs. This collapse is most noticeable as a radial shrinking of the cake. The cause of this shrinkage has generally been attributed to reduction in the viscosity of the matrix, to the point where the viscosity becomes too low to resist the flow originating in the gravity or surface tension driving forces. The temperature at which this occurs is a function of the moisture content, as well as the type of solute.

To & Flink (1978) have found that the collapse temperature can be related to water content and that this dependence is different above and below the BET monolayer value. For any given composition, the effect of water content could be calculated by the following equation :

$$\ln (T_c/T_o) = - k_2 m \qquad \text{for } m < \text{BET}$$

$$\ln (T_c/T_o) = k_3 - k_4 m \qquad \text{for } m > \text{BET}$$

where

T_c = collapse temperature
T_o = collapse temperature at m = 0
m = moisture content

During freeze-drying, when the material subjected to drying consists of an ice core, a moist layer with a moisture gradient, and a dry layer, part of the moist layer can undergo collapse if the temperature is increased beyond certain critical levels. Bellows & King (1973) studied collapse by visually observing the temperature at which the sample undergoing freeze-drying started to show puffing. They proposed that during freezing, formation of ice results in development of a concentrated amorphous solute phase (CAS), and that collapse occurs during drying when the viscosity of the CAS phase is below a critical level of 10^7 - 10^{10} cp.

Using microscopic techniques, Mackenzie (1977) has determined the collapse temperature (T_C) of a number of solutes during freeze-drying. These temperatures also show a dependence on the solutes' molecular weight. Starches and proteins have T_C values close to -10°C, while glucose and fructose are found to collapse around -40°C. Le Meste et al. (1979) reported that, for model solutions analogous to fruit juices, the influence of chemical composition on the collapse temperature during freeze-drying could not be predicted from the DSC recordings carried out on the frozen material. A very small quantity (4% of the dry weight) of polymer (pectin - protein) added to a solution of sugars and citric acid resulted in an important elevation of the T_C determined by microsocpic observation. No change could be detected on the DSC thermogram of the frozen product. For solutions containing only small solutes, T_C was lower than T_{AM}, whereas for solutions containing a polymer, T_C was close to T_{IM}. In contrast, the thermograms obtained with the freeze-dried materials showed a small increase of the glass transition temperature for the product containing pectin.

Glass transition and stability : The glass transition, which is identified on DTA-DSC scans as a shift in specific heat of the system, is a transition with which are associated several other changes in physical properties. This phenomenon has been given special attention in the field of polymer science, since the glass transition temperature is a most important characteristic in applications of polymeric materials. However, the phenomenon may be observed with similar features in any kind of glassy materials. The basic knowledge may be found in many books and reviews (e.g. Ferry, 1980; Eisenberg, 1984).

Most polymeric materials, but also small molecular weight compounds, e.g. sugars, may be cooled beyond their melting point (T_M) without crystallizing. The supercooled liquid is transformed into a glassy solid at the glass transition temperature (T_G). The most unambiguous definition of T_G is the temperature at which the thermal expansion coefficient undergoes a discontinuity. According to the free volume theory, the decrease of specific volume with temperature originates in the decrease of an interstitial volume produced by the Brownian motion of molecules or segments of molecules. T_G is the point at which the collapse of free volume can no longer occur because the molecular adjustments have been slowed by the lowering of temperature and can no longer take place within the experimental cooling times. A major feature of the glass transition is that it is a kinetic process.

The changes in mechanical properties occurring in the vicinity of T_G have a major practical interest. The modulus (e.g. Young's modulus), which has a high value in glasses, drops over a narrow temperature range around T_G. Organic glasses have then the viscoelastic behaviour typical of lightly cross-linked rubbers, or of physically entangled long chains. Depending on whether the material is cross-linked or not, the modulus exhibits a "rubbery plateau" or drops further as the temperature continues to increase.

The viscosity also shows an important decrease in the vicinity of the glass transition temperature. At temperatures well above T_M, the variation of viscosity as a function of temperature is generally Arrhenius-like. In the temperature range between T_M and T_G, Arrhenius plots show an increase in slope (equivalent to an increase in activation energy) as the temperature decreases. It is to be mentioned that the variation of viscosity as a function of temperature below T_G is still the subject of controversy.

The variation of rheological properties of polymeric materials above T_G is satisfactorily described by the so-called WLF equation (Williams, Landel & Ferry, 1955) :

$$\log a_T = \frac{C_{1G}(T - T_G)}{C_{2G} + (T - T_G)} \qquad (1)$$

where a_T is the ratio of viscosity values (or of values of another mechanical parameter) at temperatures T and T_G. C_{1G} and C_{2G} are constants, the values of which (C_{1G} = 17.4, C_{2G} = 51.6) have been shown to be nearly universal for a very wide range of materials, i.e. almost all glass-forming materials. The WLF equation has proven successful with numerous polymers and also with organic and inorganic glasses such as glucose melt (Williams <u>et al</u>., 1955). The WLF equation is valid for a temperature range usually between T_G and T_G + 100, under the provision that the material undergoes no structural change in this interval.

Water, acting as a plasticizer in most food systems, usually induces a strong decrease in the temperature of the glass transition. This effect is well-documented with biopolymers, e.g. elastin (Kakivaya & Hoeve, 1975), gelatine (Marshall & Petrie, 1980), starch (Biliaderis <u>et al</u>., 1986), hemicellulose and lignin (Kelley <u>et al</u>., 1987), as well as with small molecular weight sugars (Fig. 1) (Rasmussen & Luyet, 1969; MacKenzie & Rasmussen, 1972).

Amorphous hydrated sugars (with water content below the unfreezable water content) have been demonstrated to exhibit variations as a function of temperature of their physical properties (i.e. viscosity : Soesanto & Williams, 1981; dielectric relaxation : Chan <u>et al</u>., 1986) very similar to those observed for amorphous polymers.

According to the free volume theory of the glass transition, the effect of the plasticizing compound is similar to the effect of a temperature increase, i.e. an increase in free volume, and a weakening of intermolecular interactions (Sears & Darby, 1982).

It has been claimed by Slade & Levine (1985) that the availability of water to support chemical, physical and biological changes can be predicted by the use of a combination of the functions ΔT and ΔC where

$$\Delta T = T - T_G$$
$$\Delta C = C - C_G \text{ if } C < C_{G'}$$

or

$$\Delta T = T - T_{G'}$$
$$\Delta C = C - C_{G'} \text{ if } C > C_{G'}$$

T and C are the temperature and water content, respectively of the material under test. T_G is the glass transition temperature corresponding to the water content C_G. $C_{G'}$ is the maximum unfreezable water content in the system and $T_{G'}$ the corresponding glass transition temperature. It should be remarked that, for these authors, the temperature $T_{G'}$ is not equal to the temperature T_{G1} (Figs 1 and 2) but is a temperature between T_{AM} and T_{IM}. It has been claimed further that in the temperature range between T_G and T_M, an important increase in rates of reactions could be observed because of WLF kinetics (Slade & Levine, 1985). The WLF equation (1) indeed implies that the measured mechanical property (or related transport properties) must change in the temperature range above T_G with an apparent activation energy much higher than the activation energies for flow or diffusion which are observed in liquids at temperatures far above T_G.

For systems with a water content below the unfreezable water value, this approach seems to be interesting and deserving at least experimental investigation. A first question to be answered is the feasibility of the practical determination of T_G for food materials. DSC and DTA afford an easy determination of T_G in many materials, but the phenomenon may be broadened and then poorly detected in complex mixtures, as it is already in protein systems (Simatos _et al_., 1975; Le Meste & Duckworth, 1987). It may thus be suggested that direct measurements of mobility (translational or rotational diffusion, or mechanical measurements) may give more reliable data than microcalorimetric ones.

It is evident that the glass transition must be a critical event with regard to the evolution of physical processes in food systems. Quite a lot of experimental evidence may be provided to demonstrate the relationship between the glass transition and processes such as structural collapse of freeze-dried products (To & Flink, 1987; Le Meste _et al_., 1979), agglomeration (Downton _et al_., 1982) and crystallization of sugars (Chevalley _et al_., 1970). Referring to the features of the glass transition and of the plasticization process would certainly be an efficient approach for the prediction of the performance of stabilizing compounds (Levine & Slade, 1986) or of optimal processing parameters.

The usefulness of the concept as regards chemical or biological stability deserves further investigation. The glass transition must evidently be correlated with a change in diffusion rate. Experimental data are very scarce, particularly for food systems. From the data collected for polymers (see Crank & Park, 1968; Ferry, 1980 for reviews) the following conclusions seem to be valid. For water diffusing in various polymers such as polyvinyl-acetate, the preexponential factor D_o in the relation :

$$D = D_o \exp(-E_D/RT)$$

is several orders of magnitude larger at temperatures above T_G than below T_G. The slope of the Arrhenius plot, E_D, increases when the temperature is raised in the vicinity of T_G and then decreases. For organic vapours, the diffusion coefficient at vanishing concentration exhibits the same incurvation of the Arrhenius plot at temperatures above T_G. For many gases, on the contrary, no inflection in the Arrhenius plots was observed. This difference was attributed to an effect of the size of the diffusing molecule.

The diffusion coefficient of small foreign molecules at vanishing concentration has been compared to the mobility of the polymer as determined by mechanical measurements. Mobilities are expressed by the friction coefficients:

ξ_o = friction coefficient of the monomeric unit of the polymer
ξ_1 = friction coefficient of the foreign molecule = KT/D

For temperatures far above T_G, the friction coefficients for the chain unit and for a foreign molecule of like size are closely similar. When the temperature is closer to T_G, however, ξ_1 is considerably smaller than ξ_o. This effect is accompanied by a growing discrepancy between the apparent activation energies for diffusion and viscoelastic relaxation, the former usually being the smaller.

The various processes which are associated with the glass transition, although relying on a unique process, variation in free volume, may appear with different kinetic features, particularly different behaviours as a function of temperature or water content.

For products with water contents above the unfreezable water content, it should be remarked that melting of ice occurs in the temperature range between T_G' and T_M. The "supercooled" interstitial phase is diluted in such a way that its concentration follows the M curve. The variation of its viscosity as a function of temperature will not simply follow the WLF equation, since it will be influenced by this dilution effect. As stressed by Bellows & King (1973), the viscosity of CAS decreases when temperature increases as a consequence : (1) of the effect of temperature on viscosity (WLF equation) and (2) of the dilution. A quantitative prediction of the stability of the product as a function of temperature should take account of these combined effects and, eventually, of the effect of reactant dilution on the reaction rate.

Reactivity

Enzyme activity as affected by water activity and water content : Kertesz (1935) was one of the first to recognize that there was a "critical water concentration" for the activity of enzymes. He measured the hydrolysis of sucrose on model systems containing ground sucrose, apple pomace, and invertase mixed directly with various amounts of water, and found that (a) no reaction occurred below 4.75% moisture from -18°C to + 40°C, and (b) the rate increased with moisture content. Acker & Luck (1958) conducted a systematic study of the hydrolysis of pure lecithin added to barley malt flour and exposed to water activities (a_w) from 0.15 to 0.75 at 30°C for up to 100 days. They found that the reaction rates increased with increasing water content and that at each moisture level there was a maximum amount of product which would form. They used the extent of reaction at about 10 days as their kinetic parameter at each water activity. They noted that the onset of each reaction extrapolated down to an a_w of 0.35.

Another group of workers concerned with the mechanisms of enzyme action at low levels of hydration was that of Drapron in France. Basing their conclusions on studies in model systems containing enzymes, substrates, and a variety of water-binding agents, they developed a theory of reactions at low hydration in which the onset of the reaction is ascribed to a water level at which solvent water begins to be available in a given system (Drapron, 1972). Their experimental definitions of "solvent water" were based on the shape of the water sorption isotherms.

The above studies indicate that enzymatic reactions require a liquid phase to mobilize substrate(s) and product(s), and that as long as one exists, a reaction can begin at extremely low water activities. The issue of a potential need for protein hydration was not raised by this hypothesis. At least some of the data in the literature suggested that mobilization of the substrate(s) and product(s), may in fact occur without the presence of a liquid phase, if vapour phase transport is possible. Yagi et al. (1969) found that hydrogenase from Desulfovibrio desulfuricans could catalyze the gas-phase para-hydrogen - ortho-hydrogen conversion in the apparently totally dry state, provided that the enzyme has been stabilized by $Na_2S_2O_4$ during preparation. In a follow-up study published 10 years later, the same group (Kimura et al., 1979) confirmed the activity of this hydrogenase in the dry state for three type of reaction, i.e. conversion between para-H_2 and ortho-H_2, exchange reaction between hydrogen isotopes, and reversible oxidoreduction of an electron carrier (cytochrome C_3). Duden (1971) observed hydrolytic changes in indoxyl acetate-impregnated, freeze-dried ground vegetables (cauliflower, asparagus, leeks, green beans and wax beans) even at a_w's as low as 0.01. He concluded that the only way the indoxyl acetate could have been transported to the esterase sites within the vegetable matrix was via the vapour phase.

Recently, gas phase reactions at relatively low levels of hydration by water vapour were demonstrated by Barzana et al. (1987). Dehydrated preparations of alcohol oxidase absorbed on DEAE-cellulose vigorously catalyzed a gas-phase oxidation of ethanol vapour with molecular oxygen. The gas-phase reaction was strongly dependent on the water activity of the system. Dehydrated bienzymic catalysts, including oxidase and catalase, afforded a complete and selective conversion of the substrate to acetaldehyde. The dry alcohol oxidase was much more thermostable than in aqueous solution.

The classical studies (Acker, 1962; Drapron, 1972) have focused on diffusional limitations, on transport of substrate, product, or both. There is no doubt that water activity affects the mobility of solutes. Given the known limitations on solute mobility, it is tempting to attribute the observed limited reactions at low levels of hydration to substrate or product "immobilization" (Acker, 1962; Drapron, 1972). Caragine (1967) attempted to demonstrate that the active site of invertase was located within an envelope in the macromolecule which was accessible only by passage through a relatively narrow gap and, presumably, this was the rate-limiting factor. This concept of an intramolecular rate-limiting process, while not confirmed by Caragine's data, does have more than intuitive merit, since intramolecular diffusion appears to have been ruled out as the controlling factor in aqueous systems.

Hydration-sensitive barriers located in the immediate vicinity of the enzyme were also a possibility mentioned for invertase activity by Silver & Karel (1981). In addition, a very recent paper published by a group of Chinese workers reviews some of the theories of such protein conformation dependent barriers (Chou & Zhou, 1982). Recent studies have been quite strongly indicative of the possibility that the effect of hydration on conformation of the proteins is of major consequence in determining the effects of water activity on enzyme activity. In this respect, we wish to refer in particular to studies of Tome et al. (1978) who showed, in systems of **low viscosity**, effects of water activity which were very similar to those observed in low water activity dehydrated systems. In their studies, Tome et al. (1978) used aqueous solutions of polar liquids.

As mentioned already, our studies on the behaviour of invertase in model systems at low moisture content showed that the kinetics of sucrose hydrolysis in the low moisture region were best represented by a model in which water activity affects enzyme conformation or diffusion resistance in a resistance shell in the immediate vicinity of the enzyme molecule. This entirely phenomenologically based interpretation of the kinetics receives strong support from a recent crystallographic study by Poole & Finney (1983). These authors used direct differential IR and laser Raman spectroscopies to monitor sequential hydration of lysozyme. The authors noted conformational changes in lysozyme as a consequence of hydration to levels **lower than the monolayer value**. The largest conformational changes were noted at hydration levels just below the onset of enzyme activity. The authors state "it is tempting to suggest that this solvent-related effect is required before (enzyme) activity is possible". Baker et al. (1983) reported a hydrogen exchange study which supports the existence of hydration-related conformational changes in lysozyme.

There are other studies which support hydration-caused conformational changes affecting enzyme activity. Low & Somero (1975) studied the effect of different salt concentrations on activity of several enzymes and related the effects to the organization of water around specific functional groups. Somero & Low (1977) postulated hydration-related catalytic efficiency differences among lactate dehydrogenase homologues.

Effects of water activity on non-enzymatic browning : Among the environmental factors of storage that affect browning, temperature and water activity are of particular importance. The non-enzymatic browning reaction shows a maximum at intermediate moisture contents because of water's dual role as a solvent and as a product of the reaction. At low water activity the limiting factor is inadequate mobility; therefore, addition of water, which solubilizes or plasticizes the system, promotes the reaction. At high water contents, however, the dilution of reactants and the product inhibition of condensation reactions by water, which is a condensation product, predominate, and water strongly inhibits browning. The exact position of browning maxima depends on specific products, but generally, concentrated liquids (e.g. unfrozen fruit concentrates) and intermediate moisture foods (e.g. pie fillings and "evaporated" fruits such as prunes) are in the range of moisture content most susceptible to browning.

Very often the kinetic equations describing non-enzymatic browning in foods are complex and may require computer-aided simulation for prediction of shelf-life. On the other extreme are very simple approximations to the effects of moisture and of temperature on the browning reaction in foods. For low moisture vegetables, it was possible to correlate extensive data in the literature on discoloration with moisture content and temperature by a very simple equation (Villota et al., 1980) :

$$\ln t_f = a_o + a_1(1/T) + a_2(m - m_1)$$

where

t_f = time of failure (in days)
T = absolute temperature (^oK)
m = moisture content (g H_2O/g total weight)
m_1 = BET monolayer moisture
a_o, a_1, a_2 are constants

In the case of vegetables stored under N_2 in which colour changes were due to non-enzymatic browning and the colour changes were limiting shelf-life, values of a_o ranged from (-52) to (-35), of a_1 from (14.4×10^3) to (17.8×10^3), and of a_2 from (-0.536) to (-0.139). The values of a_1 were consistent with activation energies of 29 to 35 kcal/mol.

Effects of water on oxidative reactions in foods : The major oxidative reactions in foods are due to peroxidation of lipids and of some water-soluble vitamins. Major products of oxidation of lipids, such as fatty acids, acylglycerols and phospholipids, are the peroxides of these lipids. These peroxides can then react with the other food constituents, or they may decompose to secondary products such as aldehydes, ketones, hydroxyacids, or hydrocarbons, which are volatile and may have strong odour and off-flavour. It is these smaller, more volatile compounds that cause rancidity in foods. These secondary products as well as peroxides and lipid free radicals can react with proteins and vitamins, causing losses in nutritional value and in solubility of food constituents.

Among sensitive vitamins, vitamin C is quite important. In low pH dehydrated fruit and vegetable powders, the ascorbic acid degradation is very strongly dependent on water activity, but not on oxygen concentration. Dilute systems of similar composition, however, which have a high water activity show a pronounced oxygen dependence. The oxidation of ascorbic acid in the low water activity systems is extremely dependent on water activity. A number of studies have been conducted on this topic and they agree in this respect. The dependence is often exponential with the logarithm of the rate constant proportional to water activity, but other relations including polynomial functions have also been used to relate rate to water activity.

Lipid oxidation in foods is associated almost exclusively with unsaturated fatty acids and is often autocatalytic, with oxidation products themselves catalysing the reaction so that the rate increases with time. Water activity is a major factor affecting the rate of lipid oxidation in foods. It has long been recognized that very low water contents in fat-containing foods are conducive to rapid oxidation, and the literature documenting this phenomenon is voluminous (Karel & Yong, 1981).

Several hypotheses have been advanced to explain water's retarding effect on lipid oxidation. One factor that may be important is production of antioxidants through non-enzymatic browning. The effectiveness of products of non-enzymatic browning was demonstrated early by Griffith & Johnson (1957) and was one of the mechanisms hypothesized by Karel (1960). Several investigators in Japan and Europe have confirmed recently that intermediates in the complex set of reactions we term non-enzymatic browning are effective antioxidants. Furthermore, intermediate moisture contents would maximize the concentration of these intermediates (Eichner & Ciner-Doruk, 1981). Since high water activity promotes browning, this may be one of the explanations for the observed effects.

Studies on purified model systems containing no components capable of forming antioxidants through browning, however, also show retardation by increasing water content and the explanation for these effects is probably based on the effect of water on initiation and termination steps of the chain reaction. In purified systems, water interferes with the normal bimolecular decomposition of hydroperoxides by hydrogen bonding with amphipolar hydroperoxides formed at the lipid-water interface. In the presence of trace metals added to model systems of lyophilized emulsions, humidification at moderate levels retards oxidation because of hydration of metal ions. The reduction in rate depends on the type and hydration state of the metal salt added as well as on water content. The monomolecular rate period is primarily affected by this mechanism, although bimolecular rates are also decreased (Karel et al., 1967).

Retardation of peroxidation by increasing water content is reversed at high water activity. As water contents are increased above monolayer coverage in foods and in model systems, resistance to diffusion of solutes decreases and solubilization becomes significant. In systems containing chelating agents and antioxidants, high water contents allow solubilization of chelating agents that sequester metals as well as solubilization of antioxidants. This effect lowers the oxidation rate. However, at high water activities water may accelerate oxidation by solubilizing catalysts or by inducing swelling of macromolecules such as proteins to expose additional catalytic sites.

CONCLUSIONS

In the different processes which are of importance with regard to food quality and stability, we may find either one or, more often, both of the following major influences of water:

- Structuring effect (on molecules or on groups of molecules)
- Mobilising effects.

It may be considered that in the past too much emphasis has been given to water activity, or to "water binding" which are more or less associated with structuring effects. It is certainly appropriate that mobilizing effects which have received relatively less attention (Duckworth, 1963, 1981; Karel, 1975; Simatos et al., 1981) become subject to greater scrutiny and emphasis. The development of new methods to evaluate transport properties is highly desirable.

REFERENCES

Acker, L. (1962). Adv. Fd Res. **11**, 263.

Acker, L. & Luck, E. (1958). Z. Lebensm. Untersuch. u-Forsch. **198**, 256.

Baker, L., Hansen, J., Rao, A.M. & Bryan, W.P. (1983). Biopolymers **22**, 1637.

Barzana, E., Klibanov, A. & Karel, M. (1987). Appl. Biochem. Biotech (in press).

Bellows, R.J. & King, C.J. (1973). AiChE Symp. Ser. **132**(69), 33.

Berendsen, H. (1975). In Water, A Comprehensive Treatise, Vol. 5 (Franks, F. ed.), p. 293 Plenum Press, New York.

Bettenfeld, M.L. (1985). Thesis, University of Dijon, France.

Biliaderis, C.G., Slade, C.M., Maurice, T.J. & Juliano, B.O. (1986). J. Agric. Food Chem. **34**, 6.

Biswas, A.B., Kumsah, S.A., Pass, G. & Philips, G.O. (1975). J. Solution Chem. **4**(7), 581.

Blond, G. (1985). In Properties of Water in Foods (Simatos, D. & Multon, J.L., eds), p. 531. Martinus Nijhoff Publ., Dordrecht.

Blond, G. (1974). Personal communication.

Blond, G. (1987). To be published.

Bruin, S. & Luyben, K. (1980). Advances in Drying, Vol. 1 (Mujumdar, A., ed.), p. 155. Hemisphere Publ. Corp., Washington.

Brun, M., Quinson, J.F. & Eyraud, C. (1979). L'actualite chimique, Octobre 1979, 21.

Bull, H.B. & Breese, K. (1970). Arch. Bioch. Biophys, **137**, 299.

Busk, G.C. & Labuza, T.P. (1979). J. Food Sci. **44**(5), 1369.

Caragine, P.J. Jr. (1967). B.S. Thesis, M.I.T., Cambridge, Mass.

Carlsson, A., Lindman, B. & Nilsson, P.G. (1986). Polymer **27**, 431.

Chan, R.K., Pathmanathan, K. & Johari, G.P. (1986). J. Phys. Chem. **90**, 6358.

Chevalley, J., Rostagno, W. & Egli, R.H. (1970). Rev. Intern. Chocolat. **25**, 3.

Chou, K.C. & Zhou, G.P. (1982). J. Amer. Chem. Soc. **104**, 1409.

Colas, B. (1987). J. Dairy Res. (in press).

Crank, J. & Park, G.S. (1968). Diffusion in Polymers. Academic Press, London & New York.

Dang Vu Bien (1965). Thesis. Faculty of Science, Paris.

De Loor, G.P. & Meijboom, V.W. (1966). J. Food Technol. 1, 313.

Dote, J.L., Kivelson, D. & Schwartz, R.N. (1981). J. Phys. Chem. **85**, 2169.

Downton, G.E., Flores-Luna, J.L. & King, C.J. (1982). Indust. Eng. Chem. Fund. **21**, 447.

Drapron, R. (1972). Ann. Technol. Agric. **21**(4), 487.

Duckworth, R.B. (1963). In Recent Adv. in Food Sci., Vol. 2 (Hawthorn, J. & Leitch, J.M., eds), p. 46. Butterworths, London.

Duckworth, R.B. (1972). Proc. Inst. Fd Sci. Tech. **5**, 60.

Duckworth, R.B. (1981). In Water Activity : Influences on Food Quality (Rockland, L.B. & Stewart, G.F., eds) p. 295. Academic Press, New York.

Duckworth, R.B. (1986). Unpublished.

Duckworth, R.B., Allison, J.Y. & Clapperton, J.A. (1975). In Intermediate Moisture Foods (Davies, R., Birch, G.G. & Parker, K.J., eds), p. 89. Applied Sci. Publ., London.

Duckworth, R.B. & Kelly, C.E. (1973). J. Food Technol. **8**, 105.

Duckworth, R.B. & Smith, G.M. (1963). Proc. Nutr. Soc. **22**, 182.

Duden, R. (1971). Lebensm. Wissu. u-Technol. **4**, 205.

Eichner, K. & Ciner-Doruk, M. (1981). In Maillard Reactions in Foods (Eriksson, C., ed), p. 115. Pergamon Press, Oxford.

Eichner, K., Laible, R. & Wolf, W. (1985). In Properties of Water in Foods (Simatos, D. & Multon, J.L., eds), p. 191. Martinus Nijhoff Publ., Dordrecht.

Eisenberg, A. (1984). In Physical Properties of Polymers (Mark, J.E., Eisenberg, A., Graessley, W.W., Mandelkern, L. & Koenig, J.L., eds), p. 55. Amer. Chem. Soc., Washington, D.C.

Eisenberg, D. & McLachlan, A.D. (1986). Nature **31**, 199.

Ferry, J.D. (1980). Viscoelastic Properties of Polymers, 3rd ed. Joh5n Wiley & Sons, New York.

Fisher, L.R. & Israelachvili, J.N. (1979). Nature **277**, 548.

Flory, P.J. (1953). Principles of Polymer Chemistry. Cornell University, New York.

Flory, P.J. (1979). In Polysaccharides in Food (Blanshard, J.M.V. & Mitchell, J.R., eds), p. 51. Butterworths, London.

Franks, F. (1983). Cryobiology **20**, 335.

Franks, F. (1985a). In Properties of Water in Foods (Simatos, D. & Multon, J.L., eds), p. 497. Martinus Nijhoff Publ., Dordrecht.

Franks, F. (1985b). In Water Activity : A Credible Measure of Technological Performance and Physiological Stability. Discussion Conf., Cambridge, U.K., July 1985.

Giannakopoulos, A. & Guilbert, S. (1986). J. Food Technol. **21**, 477.

Griffin, D.M. (1981). Adv. Microb. Ecol. **5**, 91.

Griffith, T. & Johnson, J. (1957). Cereal Chem. **34**, 159.

Guerts, T.J., Walstra, P. & Mulder, H. (1974). Neth. Milk Dairy J. **28**, 46.

Guilbert, S., Giannakopoulos, A. & Cheftel, J.C. (1985). In Properties of Water in Foods (Simatos, D. & Multon, J.L., eds), p. 343. Martinus Nijohff Publ., Dordrecht.

Gurd, F.R.N. & Rothgeb, T.M. (1979). Adv. Prot. Chem **33**, 73.

Hermansson, A.M. (1986). In Functional Properties of Macromolecules (Mitchell, J.R. & Ledward, D.A.), p. 273. Elsevier Appl. Sci. Publ., London.

Hermansson, A.M. & Buchheim, W. (1981). J. Coll. Int. Sci. **81**, 519.

Hermansson, A.M. & Lucisano, M. (1982). J. Food Sci. **47**, 1955.

Iglesias, H.A. & Chirife, J. (1982). Handbook of Food Isotherms. Academic Press, New York.

Ikada, Y., Suzuki, M. & Iwata, H. (1980). In Water in Polymers (Rowland, S.P.,), p. 287. Amer. Chem. Soc., Washington, D.C.

Kakivaya, S.R. & Hoeve, C.A.J. (1975). Proc. Nat. Acad. Sci. **72**, 3505.

Karel, M. (1960). Ph.D. Thesis, M.I.T., Cambridge, Mass.

Karel, M., Fennema, O. & Lund, D. (1975). Physical Principles of Food Preservation. Marcel Dekker, New York.

Karel, M. & Flink, J.C. (1983). Adv. Drying **2**, 103.

Karel, M., Labuza, T.P. & Maloney, J.F. (1967). Cryobiology **3**, 288.

Karel, M. & Yong, S. (1981). In Water Activity : Influences on Food Quality (Rockland, L.B. & Stewart, G.F., eds), p. 511. Academic Press, New York.

Karplus, M. (1986). In Methods in Enzymology (Hirs, C.W.H. & Timasheff, S.N., eds), **181**, 283.

Katz, E.E. & Labuza, T.P. (1981). J. Food Sci. **46**, 403.

Kelley, S.S., Riabs, T.G. & Glaser, W.G. (1987). J. Materials Sci. **22**, 617.

Kertesz, Z.I. (1935). J. Amer. Chem. Soc. **57**, 1277.

Kimura, K., Suzuki, A., Inokuchi, H. & Yagi, T. (1979). Biochim. Biophys. Acta **567**, 96.

King, C.J., Carn, R.M. & Jones, R.L. (1976). J. Food Sci. **41**, 614.

Kuntz, I.D. & Kauzmann, W. (1974). Adv. Prot. Chem. **28**, 288.

Labuza, T.P. (1984). Moisture Sorption. Amer. Assoc. of Cereal Chemists, St. Paul, Minn.

Launay, B., Doublier, J.L. & Cuvelier, G. (1986). In Functional Properties of Food Macromolecules (Mitchell, J.R. & Ledward, D.A., eds), p. 1. Elsevier Appl. Sci. Publ., London.

Le Meste, M. (1986). To be published.

Le Meste, M., Diallo, F. & Simatos, D. (1979). Proc. XVth Intern. Congress Refrig., Vol 3, 261.

Le Meste, M. & Duckworth, R.B. (1987). To be published.

Le Meste, M. & Voilley, A. (1987). J. Phys. Chem. (in press).

Levine, H. & Slade, L. (1986). Carbohyd. Polymer **6**, 213.

Lillford, P.J., Clark, A.H. & Jones, D.V. (1980). In Water in Polymers (Rowland, S.P., ed). A.C.S. Symp. Series 177, Amer. Chem. Soc., Washington, D.C.

Loncin, M. (1980). Food Process Engineering (Linko, P., Malkki, Y., Olkku, J. & Larinkari, J., eds), p. 364. Elsevier Appl. Sc. Publ., London.

Low, P.S. & Somero, G.N. (1975). Proc. Natl. Acad. Sci. USA **72**, 3305.

McClung, R.E.D. & Kivelson, D. (1968). J. Chem. Phys. **49**, 3380.

Mackay, D.H.J. & Wilson, K.R. (1986). J. Biomolecular Structure and Dynamics **4**, 491.

MacKenzie, A.P. (1977). Phil. Trans. Roy. Soc. Lond. **B278**, 167.

MacKenzie, A.P. & Rasmussen, D.H. (1972). In Water Structure at the Polymer Interface (Jellinek, H.H.G, ed.), p. 146. Plenum Press, N.Y.

Maquet, J., Theveneau, H., Djabourov, M., Leblond, J. & Papon, P. (1986). Polymer **2**, 1103.

Marshall, A.S. & Petrie, S.E.B. (1980). J. Photogr. Sci. **28**, 128.

Maurice, T.J., Slade, L., Sirett, R.R. & Page, C.M. (1985). In Properties of Water in Foods (Simatos, D. & Multon, J.L., eds), p. 211. Martinus Nijhoff Publ., Dordrecht.

Menting, L.C., Hoogstad, B. & Thijssen, H.A.C. (1970). J. Food Technol. **5**, 111.

Mitchell, J.R. (1979). In Polysaccharides in Food (Blanshard, J.M.V. & Mitchell, J.R., eds), p. 51. Butterworths, London.

Mocquot, G. (1947). Lait **27**, 576.

Moreira, T. (1976). Thesis, Universite de Dijon.

Morozov, V.N. & Gevorkian, S.G. (1985). Biopolymers **24**, 1785.

Muhr, A.H. & Blanshard, J.M.V. (1986). J. Food Technol. **21**, 587 & 683.

Poole, P.L. & Finney, J.L. (1983). Biopolymers **22**, 255.

Rasmussen, D. & Luyet, B. (1969). Biodynamics **10**, 319.

Rha, C.K. & Pradipasena, P. (1986). In Functional Properties of Food Macromolecules (Mitchell, J.R. & Ledward, D.A., eds), p. 79. Elsevier Appl. Sci. Publ., London.

Ross, K.D. (1978). J. Food Sci. **43**, 1812.

Ruegg, M., Luscher, M. & Blanc, G. (1974). J. Dairy Sci. **57**, 387.

Rulkens, W.H. & Thijssen, H.A.C. (1972). J. Food Tech. **7**, 79.

Rupley, J.A., Gratton, E. & Careri, G. (1983). Trends Biochem. Sci. **8**(1), 18.

Sears, J.K. & Darby, J.R. (1982). The Technology of Plasticizers. John Wiley Inters. Publ., New York.

Silver, M. & Karel, M. (1981). J. Food Biochem. **5**, 283.

Simatos, D., Faure, M., Bonjour, E. & Couach, M. (1975). Cryobiology **12**, 202.

Simatos, D., Faure, M., Bonjour, E. & Couach, M. (1975). In Water Relations of Foods (Duckworth, R.B., ed), p. 193. Academic Press, London.

Simatos, D., Le Meste, M., Petroff, D. & Halphen, B. (1981). In Water Activity : Influences on Food Quality (Rockland, L.B. & Stewart, G.F., eds), p. 319. Academic Press, New York.

Slade, L. & Levine, H. (1985). In Water Activity : A Credible Measure of Technological Performance and Physiological Viability? Discussion Conf., Cambridge, U.K.

Small, D.M. (1986). The Physical Chemistry of Lipids. Plenum Press, New York.

Soesanto, T. & Williams, M.C. (1981). J. Phys. Chem. **85**, 3338.

Somero, G.N. & Low, P.S. (1977). Nature **266**, 276.

Stein, D.L. (1986). In Structure, Dynamics and Function of Biomolecules (Ehrenberg, A., Rigler, R., Graslund, A. & Nilsson, L., eds), p. 70. Springer Verlag.

Suggett, A. (1975). In Water Relations of Foods (Duckworth, R.B., ed.), p. 23. Academic Press, London.

Tanaka, T. (1981). Sc. Am **244**(1), 110.

Thijssen, H.A.C. 91979). Lebensm. Wiss. Technol. **12**, 308.

Thijssen, H.A.C. & Rulkens, W.H. (1968). De Ingenieur **80**(47), 45.

To, E.C. & Flink, J.M. (1978). J. Food Technol. **13**, 551.

Toei, R. (1986). In Drying '86, Vol. 2 (Mujumdar, A., ed.), p. 880. Hemisphere Publ. Corp., Washington.

Tome, D., Nicolas, J. & Drapron, R. (1978). Lebensm. Wiss. u-Technol. **11**, 38.

Tsouroflis, S., Flink, J.M. & Karel, M. (1976). J. Sci. Food Agric. **27**, 509.

Van Gusteren, W.F. & Karplus, M. (1981). Nature **293**, 677.

van den Berg, C. (1986). In Concentration and Drying of Foods (MacCarthy, D., ed.), p. 11. Elsevier Appl. Sci. Publ., London.

Villota, R., Saguy, I. & Karel, M. (1980). J. Food Sci. **45**, 398.

Voilley, A. & Le Meste, M. (1985). In Properties of Water in Foods (Simatos, E. & Multon, J.L., eds), p. 357. Martinus Nijohff Publ., Dordrecht.

Voilley, A. & Roques, M. (1987). In Physical Properties of Foods **2**, 109.

Walstra, P. (1973). Kolloid Z. Polymere **251**, 603.

Wilke, C.R. & Chang, P. (1955). Amer. Inst. Chem. Eng. J. **1**, 264.

Williams, M.L., Landel, R.F. & Ferry, J.D. (1955). J. Amer. Chem. Soc. **77**, 3701.

Wolfenden, R., Andersson, L., Cullis, P.M. & Southgate, C.C.B. (1981). Biochem. **20**, 849.

Yagi, T., Tsoda, M., Mori, Y. & Inokuchi, H. (1969). J. Amer. Chem. Soc. **91**, 2801.

Young, F.E. & Jones, F.T. (1949). J. Phys. Coll. Chem. **53**, 1334.

CHARACTERIZATION OF THE STATE OF WATER IN FOODS - BIOLOGICAL ASPECTS

G.W. GOULD
Unilever Research Laboratory
Colworth House, Sharnbrook
Bedford, MK44 1LQ, U.K.

AND

J.H.B. CHRISTIAN
CSIRO Division of Food Research
P.O. Box 52, North Ryde,
NSW 2113, Australia

INTRODUCTION

The majority of food preservation procedures have a long history of traditional use. Most of them were derived empirically, and preservation through modification of the state of water in foods is no exception. Originally achieved by simply drying foods in the sun, it must have been one of the earliest procedures utilised by man. Along with most of the other economically and socially important food preservation methods that are in current use, it is surprising that so little is still known about its fundamental mode of action. However, with regard to water status, a major conceptual advance occurred when Scott (1957), building predominantly on previous observations of Mossel & Westerdijk (1949) and Scott (1953), proposed that, as far as effects on the physiology of food spoilage and food poisoning microorganisms were concerned, the thermodynamic water activity (a_w) was a key determinant, to a large extent independent of the means by which the particular value was obtained. In particular, the growth limits for different types of microorganisms have since been commonly described in this way, with the 'a_w limits for growth' quoted for each species or strain, below which growth cannot occur. At the same time, the improved availability of easy-to-use electric hygrometers and other devices that allow reasonably rapid and accurate estimation of the a_w values of foods has encouraged the use of this parameter. Different types of food spoilage and food poisoning microorganisms vary greatly in their ability to grow and to metabolise as the a_w is reduced. Therefore, like Eh, pH value and temperature, a_w has been increasingly promoted as a practically-useful determinant of microbial activity (Troller & Christian, 1978). At present, however, it is becoming increasingly evident that there are other water-related factors that are not necessarily a_w-dependent, but that can also greatly influence that stability of foods. This paper therefore attempts to highlight the biological significance that will emerge from an improved understanding of these new factors.

WATER ACTIVITY AS A DETERMINANT OF GROWTH

There is no doubt that the increased use of a_w as a determinant of microbial growth and metabolic activity, as well as of the activity of food-deteriorative enzymic and chemical reactions, has been of great practical value. And this has been true despite the fact that, at the level of cell biochemistry, physiology and metabolism, there are no obvious fundamental reasons why a_w should provide such a soundly-based determinant. An examination of the reactions of different microorganisms to lowered a_w will make this clear.

First, it is important to appreciate that, being prokaryotic, all bacteria are morphologically simple, small, uncompartmented, usually unicellular, and therefore always in intimate contact with their environment. Eukaryotic microorganisms of spoilage and public health significance, such as yeasts and moulds, have more complex cells, are more often multicellular, but are still small. The smallness and the absence of substantially water-impermeable barriers ensure that microorganisms in general must tend to come into rapid osmotic equilibrium with their surroundings. Consequently, if the environmental a_w is reduced, microbial cells tend to lose water as osmotic equilibrium across the cell membrane is reestablished.

In general, it has been observed that if the reduction in a_w is brought about by a rise in the concentration of environmental solutes that are unable to readily penetrate the cell membrane, vegetative microorganisms react by raising the levels of their internal solutes, by synthesis or by uptake from the medium (Christian, 1955), sufficiently to just exceed the external osmolality and, in this way, avoid excessive loss of cytoplasmic water (Brown & Simpson, 1972; Gould & Measures, 1977). They thus appear to operate a 'homeostatic mechanism' with respect to cell water content.

It is widely agreed that the flow of water across the cell membranes of microorganisms is passive. The homeostatic control of cell water content must therefore be achieved as a consequence of such changes in cytoplasmic osmolality and/or as a result of changes in physical forces, for instance such as may result from changes in tension, or contraction or expansion of the cell membrane or wall. The major control, in vegetative cells, is certainly exerted through the manipulation of key cytoplasmic solutes, and this is brought about via changes in metabolic pathways within the cell and by changes in the activities of membrane transport systems.

Many of the so-called 'osmoregulatory' solutes share the property of interfering minimally with the activities of intracellular enzymes at concentrations at which most of the common environmental solutes, in particular sodium chloride, cause severe inhibition. The intracellular solutes have therefore been termed 'compatible' solutes (Brown, 1978). They also share the property of high solubility, which is presumably a prerequisite for the evolutionary selection of an osmoregulatory solute.

Indeed, some of them may rise within the cytoplasm of metabolising and growing cells to concentrations well above molar. Chemically, however, they are very diverse, including amino acids (e.g. glutamic acid; proline) and derivatives of amino acids (e.g. glutamine; trimethyl glycine or betaine), cations (e.g. potassium), sugars (e.g. glucose; sucrose) and polyols (e.g. glycerol; arabitol), see Gould (1985).

It is important to remember that, in addition to being specific to a particular organism, the particular intracellular solute that is accumulated often depends greatly on the medium surrounding the osmoregulating cell. For example, the intracellular glutamic acid concentration of *Pseudomonas fluorescens* growing in NaCl-adjusted medium at a_w 0.98 rose 23-fold to become 90% of the amino acid pool, but when a similar a_w was achieved with sorbitol, no such rise in glutamic acid occurred, but sorbitol simply accumulated intracellularly to about twice the external concentration (Prior *et al*., 1987).

An additional phenomenon, the significance of which has yet to be fully worked out, is that, in *E. coli*, high osmolarities of salt or sucrose bring about a fall in the levels of certain 'membrane-derived oligosaccharides' that are located in periplasmic space (Kennedy, 1982, 1984; Bohin & Kennedy, 1984). It is believed that, at low osmolarities, these anionic polymers, and the positively-charged counterions that must be associated with them in order to preserve electrical neutrality, act to maintain osmotic pressure in the periplasm, and therefore around the enclosed cytoplasmic compartment.

The framework of the general osmoregulatory system described above makes it easy to appreciate why a_w is a useful determinant of growth. Most solutes of importance in foods are non-penetrant in the sense that they do not rapidly, passively, pass across cell membranes. Consequently, microorganisms in foods will generally be operating at some level of the osmoregulatory response, and a determination of a_w will give a good indication of the extent to which this response is occurring. However, many factors may interfere with this simple relationship. In particular, there are general and specific effects related to different environmental solutes that may lead to substantial departures from the relationship.

GENERAL AND SPECIFIC SOLUTE EFFECTS

There is no single guide or principle governing the solute effects one should expect with a particular microorganism. In some instances, for example, sodium chloride will inhibit more effectively than glycerol at equivalent osmolality (e.g. *Escherichia coli*); in other instances the converse is true (e.g. *Staphlylococcus aureus*; Gould & Measures, 1977). With less often-studied solutes the effects may be unexpectedly great (Table 1). For example, Vaamonde *et al*. (1986) found *S. aureus*, incubated at 30°C, to be inhibited at about a_w 0.86 by sodium acetate; at about 0.90 by sodium lactate, at about 0.93 by xylitol and at an a_w as high as 0.95 by erythritol, and there is no obvious explanation of these differences.

Of particular importance is the well-documented general effect that an increase in the concentrations of environmental solutes (such as glycerol) that are able to rapidly equilibrate across the cell membrane, and therefore do not cause a major loss of cell water and contraction of

TABLE 1
Specific solute effects on growth-inhibitory a_w values

Organism	Solute	Approx. a_w + at which growth is prevented	Ref. *
Pseudomonas fluorescens	NaCl	0.97	1
	glycerol	0.95	1
Clostridium novyi	NaCl	0.95	2
	glucose	0.95	2
	glycerol	0.935	2
Staphylococcus aureus	erythritol	0.95	3
	xylitol	0.93	3
	glycerol	0.89	1
	sucrose	0.87	4
	sodium acetate	0.86	3
	NaCl	0.85	1
Clostridium botulinum A	NaCl	0.94	5
	glycerol	0.92	5
Bacillus cereus	NaCl	0.92	1
	glycerol	0.92	1
Bacillus subtilis	glycerol	0.92	1
	NaCl	0.90	1
Basipetospora halophila	glycerol	0.76	6
	glycerol + fructose	0.75	6
	NaCl	0.73	6
Eurotium species	NaCl	0.73	6
	glycerol	0.70	6
	glucose + fructose	0.66	6

\+ Figures listed are approximate a_w values at which growth does not occur under otherwise near-optimal conditions. (Generally, growth was shown to occur at a_w values about 0.01 units higher than the inhibitory levels).

* References: (1) Gould & Measures (1977); (2) Jakobsen (1985); (3) Vaamonde et al. (1986); (4) Christian (1981); (5) Baird-Parker & Freame (1967); (6) Andrews & Pitt (1987).

the cell cytoplasm ('plasmolysis'), do not elicit the characteristic response seen with substantially non-penetrant solutes. (Anagnostopoulos & Dhavises, 1980, 1982; Kroll & Anagnostopoulos, 1981). 'Osmoregulatory pathways' for solutes such as glutamate, proline, betaine (Landfald & Strom, 1986) are all stimulated, however, by a range of non-penetrant solutes. This has been taken to indicate that the osmoregulation response is triggered by some change related to the loss of cell water rather than to the presence of solute molecules, the change in osmolality, or the reduction in a_w *per se*. One possibility, for which there is supporting evidence, is that the increase in concentration of particular cytoplasmic constituents, that results directly from plasmolysis, are the key biochemical triggers for osmoregulation. For instance, in some microorganisms a rise in the intracellular concentration of potassium, i.e. simply resulting from loss of water or via increased uptake (Christian & Waltho, 1964), will increase the relative activities of enzymes on the pathway leading to the synthesis of glutamic acid, γ-amino butyric acid and proline, which are osmoregulatory solutes in many organisms and rise in concentration in some drought-, saline- and frost-resistant plants and in some estuarine animals that are exposed to osmotic stress (Gould & Measures, 1977). On the other hand, other studies have shown that in some microorganisms (e.g. *Rhizobium*; Hua *et al*., 1982), direct potassium stimulation of intermediary metabolism does not operate.

Modern genetic analyses (Csonka, 1982; Le Rudelier *et al*., 1984) have helped to clarify the details of the osmoregulatory response at the molecular level. There are about twelve genes involved in osmoregulation in Gram-negative bacteria such as *Salmonella* and *Escherichia coli*. Of particular importance, raised cytoplasmic concentrations of potassium have been shown to directly activate the pro U gene in *Salmonella typhimurium* (Sutherland *et al*., 1986; Higgins *et al*., 1987), and this gene codes for the periplasmic binding protein for betaine (which is a more effective osmoprotectant than proline; Milner *et al*., 1987), which is known to be induced by high osmolality obtained with non-penetrant solutes. (Cells starved of potassium cannot, therefore, effectively osmoregulate). It has recently been shown (Higgins, 1988) that the rise in potassium brings about a change in DNA supercoiling and it is this that causes the specific induction of pro U.

It is most likely that the initial trigger has a predominantly physical basis and is dependant on the reduction in turgor in the cell membrane that accompanies loss of cell water, and that this modulates the activity of porins and transport proteins, e.g. for potassium, in the cell membrane (Le Rudelier *et al*., 1984). Studies of the potassium transport system by Laimins (1980) have clearly shown a_w-dependant alterations in 'Kdp' potassium transport protein conformation and synthesis in *E.coli*. An attraction of this hypothesis is that, in the successfully osmoregulating cell, turgor would increase as the cytoplasm rehydrated, and offer an explanation of the accurate 'fine tuning' of cell solute levels to match the osmolality of the environment that is actually observed.

The likely basic osmoregulatory sequence is therefore as follows: loss of turgor results in physical changes in the membrane which raise the rate of potassium accumulation which, in turn, activates genes whose products are concerned with uptake and/or synthesis of 'osmoregulatory' or 'compatible' solutes. At the same time, changes in membrane lipids commence so that salt-adapted cells have a higher proportion of negatively-charged lipids in their membranes than unadapted cells (Russell _et al._, 1985; Adam _et al._, 1987).

Interference with the osmoregulatory homeostatic mechanism is an obvious objective in food preservation technology. Since osmoregulation is an energy-requiring process, interference can be achieved by restricting the energy available to the cell, and indeed this is done in a crude way, e.g. by oxygen-free packaging or other means (Gould _et al._, 1983). **More selective interference with the sequence should be possible as more is learnt about its biochemical basis.**

KINETIC IMPEDIMENTS TO METABOLISM AND MICROBIAL GROWTH - HIGH VISCOSITY AND GLASSY STATES

Glasses are metastable systems that are supersaturated or supercooled, and characterised by very high viscosity (Kauzmann, 1984; Fig.1). As a consequence of this, all reactions that depend on molecular diffusion are essentially prevented (Franks, 1985). They therefore delay chemical and enzymic changes in food systems. It is also believed that they may play an important role in the stability and longevity of some organisms that resist relatively dry conditions for long periods of time. Their effects may, in part, be due to restriction of diffusion and, in part, to the interface between glass and intracellular macromolecules being similar to the macromolecule interface with a more dilute solution (Burke, 1986), and therefore protective of molecular structure.

Glasses will characteristically have lower water vapour pressures than the corresponding crystalline solids, and therefore resist further dehydration. Some glasses remain (meta-) stable at temperatures even above about 90°C, with levels of water of 10 to 20 g per 100 g total weight or so, and may therefore clearly add to the protection of dehydrated foods from chemical and physical deterioration. One would predict that high viscosity states would greatly interfere with, and glassy states prevent, the growth of microorganisms in foods, for instance through the restriction of diffusion of nutrients to, and end-products away from, this is so (Slade & Levine, 1985). This clearly an area where much remains to be done in order to identify the potential for deliberate use of high viscosity states, and bearing in mind that **conventional a_w determination will not define these metastable states.**

ANHYDROBIOSIS

The majority of living organisms are inactivated by desiccation. However, there are some microorganisms and also multicellular eukaryotes that exhibit extreme tolerance and are able to survive almost complete desiccation for long periods of time (Crowe & Clegg, 1973; 1978). In most instances, however, survivability is maximal only if the dehydration occurs slowly, i.e. an 'adaptation period' is required. Much research has therefore been undertaken to determine the nature of this adaptation.

An early observation was that solutes such as glycerol and trehalose commonly accumulated in desiccation-tolerant organisms. Although the presence of these solutes must reduce the rates of water loss from drying organisms, some of them additionally, and more importantly, protect the

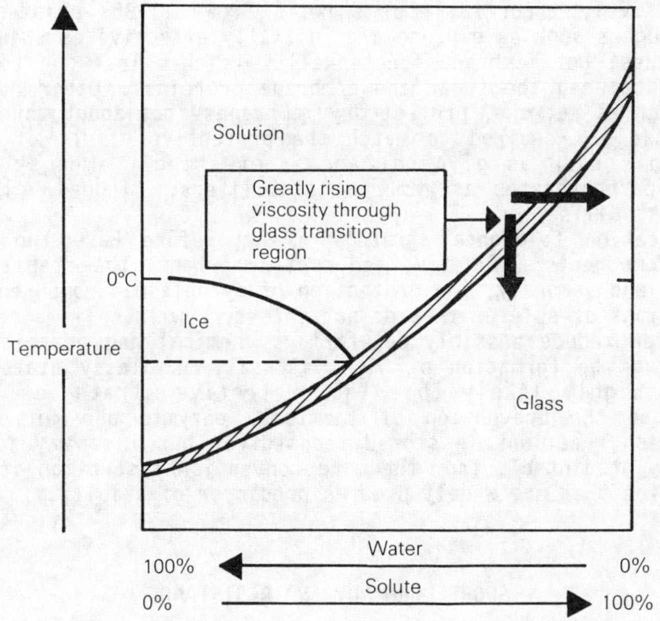

Figure 1. Diagrammatic representation of the relationship of the glass transition region to solute concentration and temperature. The broad arrows indicate the direction of sharply rising viscosity (From diagrams in Mackenzie, 1977; Slade & Levine, 1985; Burke, 1986).

viability of the completely dried cells. There is abundant evidence now that the key element in this protection is the ability of some of the polyhydroxy solutes to prevent major changes occurring in the structure of membranes as water is removed from them. The mechanism seems to involve the substitution of the carbohydrates, and trehalose in particular, for hydrogen-bonded water as the water is lost. This is the so-called 'water replacement hypothesis' (Crowe, 1971; Clegg, 1974; Crowe et al., 1984). Molecular modelling studies have indicated a water replacement role in which one trehalose molecule interacts with three phospholipid head groups (Gaber et al., 1986). There is good evidence from physical studies that unprotected membranes undergo major structural changes on desiccation (Crowe & Crowe, 1986) and, consequently, massive leakage of cytoplasmic contents results during rehydration. Protected membranes rehydrate with minimal structural interference and therefore permit minimal leakage.

Carbohydrates other than trehalose similarly stabilise membranes but less effectively. For example, Crowe & Crowe (1986) pointed out that reducing sugars such as glucose are initially effective as stabilisers of dry membranes, but membrane functionality is then lost due to 'browning reactions' between the sugar and membrane proteins. Other non-reducing sugars, such as sucrose, protect dry membranes, but about three times as much sucrose is required to match the protective effect of trehalose. Polyalcohols such as glycerol are interestingly amongst the least effective carbohydrates as membrane stabilisers. Indeed, glycerol may promote destabilisation.

Desiccation tolerance strategy may therefore have two major and distinct components: firstly, and most important, the stabilisation of membranes; and secondly, the protection of cytoplasmic components by high concentrations of solutes that do not interfere with cell macromolecules, and may even reduce possibly deleterious chemical and enzymic reactions by promoting the formation of high viscosity or glassy states. It is therefore highly likely that the protective strategies will have relevance to the prevention of chemical, enzymic and microbiological deterioration reactions in stored foodstuffs, **but in a way that is not necessarily predictable from the more conventional sorption isotherm/a_w-relationships that are widely used as predictor of stability.**

SPORE DORMANCY AND RESISTANCE

If the concentration of non-penetrant solutes around a vegetative microorganism is so high as to overwhelm its capacity to osmoregulate, then water is irreversibly lost. The partly dehydrated cell stops growing, becomes metabolically inert, and sometimes survives in this state for long periods of time. In addition, and of some practical importance, such osmotically-dehydrated cells may increase greatly in resistance to heat, even by factors in excess of 1,000-fold (e.g. Salmonella; Corry, 1974 and yeasts; Corry, 1976). In these respects, such cells superficially resemble bacterial endospores which are by far the most dormant and resistant forms of life on earth. Interestingly, sound data have recently been obtained indicating that the central cytoplasmic contents of endospores, **even when these cells are suspended in pure water**, maintain a remarkable level of dehydration, even though the **total** water content of the spore may be high. For some species these data indicate that the water content of the spore cytoplasm is as low as about 20 g per 100 g wet weight, which may be low enough to largely account for the enormous heat resistance and the extremely low metabolic activity of these cells (Beaman et al., 1982; Tisa et al., 1982). The most resistant spores tend to have the smallest, most condensed, cytoplasmic compartments and the least resistant spores have the largest and least condensed cytoplasmic compartments (Bayliss et al., 1981; Beaman et al., 1982, 1984).

There is a number of hypotheses concerning the mechanisms by which the necessary degree of dehydration is first achieved during spore formation and then maintained for long periods of time thereafter. So far, the detailed mechanisms, at the molecular level, have not been elucidated, but the hypotheses all centre on the fact that the spore is compartmentalised in a manner that is unusual in a prokaryotic cell. Lewis et al. (1960) first suggested that the cytoplasm in the central compartment of the spore may be dehydrated as a result of **contraction** of electronegative peptidoglycan, cross-linked by calcium ions, in the cortex compartment that surrounds it. The opposite, that **expansion** of peptidoglycan in the cortex by electrostatic repulsion of unshielded negatively-charged carboxyl groups brought about dehydration of the enclosed cytoplasm, was proposed by Alderton & Snell (1963). Gould & Dring (1975) reasoned that the electronegative cortex, and positively charged counterions associated with it, acted as an osmoregulatory organelle that maintained a low water content in the enclosed cytoplasm as long as this did not contain high levels of small molecules in solution (and therefore osmotically active). Warth (1985) reasoned that, if resistance resulted from the lowered cytoplasmic water content alone, then dehydration equivalent to a reduction in a_w of about 0.25 was required and, for this, generation of pressure by the cortex on the enclosed cytoplasmic compartment was necessary, e.g. by anisotropic expansion of peptidoglycan in the cortex. Similarly, Algie et al. (1983) calculated that the probable tensile strength of peptidoglycan was compatible with the generation of pressure on the enclosed cytoplasm sufficient to maintain an 'effective a_w' within it of near to 0.8, essentially by reverse osmosis.

Although the detailed mechanism, at the molecular level, is therefore still uncertain, there is ample evidence for the low water status of the spore cytoplasm and for its importance in dormancy and resistance. Clearly, the more rigidly a macromolecule is held, the more resistant it will be to denaturation by unfolding. Indeed, increase in resistance to heat to the extent of 10^5-fold have been reported for enzymes immobilised in polymethacrylate gels **with water contents as high as 50% and presumably at high water activities** (Martinek et al., 1977). It is the immobilisation and high viscosity in such systems that are the determinants of resistance rather than the water contents or water activities per se (Careri, 1982).

A major component of the dormancy-resistance mechanism of bacterial endospores is, therefore, most likely to be the relative dehydration of the central cytoplasm, maintained by the surrounding cortex (Fig. 2). However, the cortex itself may be relatively hydrated so that the **total** water content of the spore is not low, ranging up to about 70 g water per 100 g wet weight in some organisms (Beaman et al., 1982). In this sense, the spore represents a very unusual system in which the key sensitive parts are relatively dehydrated, and therefore very well preserved, whilst the inherently stable parts (in the cortex), are relatively hydrated. **A similarly effective total preservation of foodstuffs, at 70% total water content, would be very attractive should we ever be able to copy the spore.**

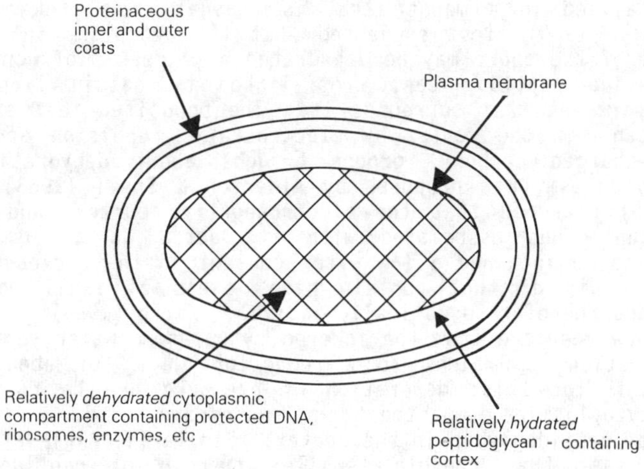

Figure 2. Structure and water relations of a typical bacterial endospore.

CONCLUSIONS

Whatever the detailed mechanism, it is the osmoregulatory capacity of a particular cell that determines, to a large extent, its osmotolerance in most practically-important situations because the environmental solutes that are most often present in foods fall into the class that do not readily penetrate the cell membrane and are therefore able to effectively plasmolyse cells. Consequently, any measurement that correlates closely with osmolality in a particular substrate will normally give a good indication of the potential for growth and metabolic activity of particular groups of food spoilage and food poisoning microorganisms. It is really for this reason that direct or indirect measurements that allow rapid estimation of, for example, equilibrium relative humidity, water activity, depression of freezing point, etc. have found increasing use. Indeed, such measurements have become so practically useful that they have entered into legislation, e.g. for the categorisation of foods for international trade with respect to their potential for supporting the growth of spoilage and food poisoning microorganisms.

Of course, there never was any reason to suppose that cells have evolved mechanisms for sensing water activity *per se*. One may therefore justifiably ask whether there is some other parameter, or set of parameters, that would form the basis of more rationally-based criteria of cell activity as influenced by water and the aqueous environment. At present, there is not. However, one may speculate that, based on current understanding, such criteria should certainly include some factor related to environmental osmolality, but **also** some measure of membrane permeability to the major solutes that are present, **plus** some factor related to the ionic versus non-ionic nature of the solutes, **plu** some factor to take account of any specific nutritional or toxic effects of the molecules that are present.

In addition, the thermodynamic water activity is an equilibrium function and therefore takes no account of dynamic factors. Consequently, there is a danger that we may miss phenomena, and opportunities, that have a **kinetic** rather than an **equilibrium** basis, such as those that may derive from better directed control of viscosity/diffusivity in foods or from more soundly-based exploitation of metastable glassy states. So far, a synthesis of all these factors has not been undertaken.

We should, arguably, derive new approaches from studies of natural phenomena. For instance the food technologist should learn new practically useful approaches from the obvious key importance, in nature, of membrane stabilisation in desiccation-tolerant organisms. And water compartmentalisation on the micro-scale, as practised by the bacterial spore, would be a difficult, but exciting, target for practical exploitation.

In the meantime, we must recognise that the concept of 'water activity' has been valuable in physiological studies of microorganisms and other cells principally because measured values generally correlate well with the potential for growth and metabolic activity. And, as indicated above, there are often soundly-based reasons why a good correlation should exist. There are also exceptions, however, where the correlation is not strong. In some such instances the reasons for this are well understood, so that, in fact, a good correlation should not be expected. In other instances the reasons for a poor correlation are not understood. However, these exceptions are not good reasons for abandoning the a_w concept in the present state of knowledge of alternative explanations. Indeed, there are clear analogies with the concept of pH, which remains of immense utility in predicting the microbial stability of foods although, as with a_w, there are many additional and interacting factors that influence the pH tolerance of microorganisms. For example, many organic acids invaluable in food preservation have inhibitory effects far greater than those predicted from pH values alone.

Overall, therefore, to some extent because it **has** been found to be so practically useful, the uncritical use of water activity as a determinant of cell activity has probably had the unfortunate effect of steering attention away from some of the more fundamental aspects of the water relations of cells. However, the examples given in this paper illustrate that some exciting possibilities exist for the **next** phase of 'biological aspects of water in foods' research, when the already-so-useful a_w/stability concept is further developed, perhaps to take in some of the phenomena that nature has had time to perfect, over millions of years of evolution.

REFERENCES

Adams, R., Bygraves, J., Kogut, M. & Russell, N.J. (1987). J. gen. Microbiol. **133**, 1861.

Alderton, G. & Snell, N.S. (1963). Biochem. Biophys. Res. Comm. **10**, 139.

Anagnostopoulos, G.D. & Dhavises, G. (1980). In Microbial Growth in Extremes of Environment. Soc. Appl. Bact. Tech. Ser. No. 15 (Gould, G.W. & Corry, J.E.L., eds), p. 141. Academic Press, London.

Anagnostopoulos, G.D. & Dhavises, G. (1982). J. appl. Bact. **53**, 173.

Andrews, S. & Pitt, J.I. (1987). J. gen. Microbiol. **133**, 233.

Baird-Parker, A.C. & Freame, B.C. (1967). J. appl. Bact. **30**, 420.

Bayliss, C.E., Waites, W.M. & King, N.R. (1981). J. appl. Bact. **50**, 379.

Beaman, T.C., Greenamyrem, J., Corner, T., Pankrantz, H. & Gerhardt, P. (1982). J. Bact. **150**, 870.

Beaman, T.C., Koshikawa, T., Pankrantz, H.S. & Gerhardt, P. (1984). FEMS Microbiol. Lett. **24**, 47.

Bohin, J.P. & Kennedy, E.P. (1984). J. Bact. **157**, 956.

Brown, A.D. (1978). Adv. Microb. Physiol. **17**, 181.

Brown, A.D. & Simpson, J.R. (1972). J. gen. Microbiol. **72**,589.

Burke, M.J. (1986). In Membranes, Metabolism, and Dry Organisms (Leopold, A.C., ed), p. 358. Cornell University Press, Ithaca, New York.

Careri, G. (1982). In Biophysics of Water (Franks, F. & Mathias, S.F. eds), p. 58. John Wiley, Chichester.

Christian, J.H.B. (1955). Aust. J. Biol. Sci. **8**, 490.

Christian, J.H.B. (1981). In Water Activity : Influences on Food Quality (Rockland, L.B. & Stewart, G.F., eds), p. 825. Academic Press, New York.

Christian, J.H.B. & Waltho, J.A. (1964). J. gen. Microbiol. **25**, 97.

Clegg, J.S. (1974). Trans. Amer. Micros. Soc. **93**, 481.

Corry, J.E.L. (1974). J. appl. Bact. **37**, 31.

Corry, J.E.L. (1976). J. appl. Bact. **40**, 269.

Crowe, J.H. (1971). Amer. Naturalist **105**, 563.

Crowe, J.H. & Clegg, J.S. (1973). Anhydrobiosis. Dowden, Hutchinson & Ross, Stroudsburg, PA.

Crowe, J.H. & Clegg, J.S. (1978). Dry Biological Systems. Academic Press, New York.

Crowe, J.H. & Crowe, L.M. (1986 a). In Membranes, Metabolism and Dry Organisms. (Leopold, A.C., ed.), p. 188. Cornell University Press, Ithaca, New York.

Crowe, J.H., Crowe, L.M. & Chapman, D. (1984). Science, N.Y. **223**, 701.

Crowe, L.M. & Crowe, J.H. (1986 b). In Membranes, Metabolism and Dry Organisms. (Leopold, A.C., ed.), p. 210. Cornell University Press, Ithaca, N.Y.

Csonka, L.N. (1981). Mol. Gen. Genet. **182**, 82.

Franks, F. (1985). Biophysics and Biochemistry at Low Temperatures. Cambridge University Press, Cambridge.

Gaber, B.P., Chandresekhar, I. & Pattabiraman, N. (1986). In Membranes, Metabolism and Dry Organisms. (Leopold, A.C., ed.), p. 231. Cornell University Press, Ithaca, N.Y.

Gould, G.W., Brown, M.H. & Fletcher, B.C. (1983). In Food Microbiology: Advances and Prospects (Roberts, T.A. & Skinner, F.A., eds), p. 67. Academic Press, London.

Gould, G.W. (1985). In Properties of Water in Foods (Simatos, D. & Multon, J.L., eds), p. 229. Martinus Nijhoff, Dordrecht, The Netherlands.

Gould, G.W. & Dring, G.J. (1975). Nature, Lond. **258**, 402.

Gould, G.W. & Measures, J.C. (1977). Phil. Trans. Roy. Soc. Lond. B. **278**, 151.

Higgins, C.F. (1988). In Homeostatic Mechanisms in Microorganisms (Banks, J.G., Board, R.G. & Gould, G.W., eds), in press. Bath University Press, Bath, U.K.

Higgins, C.F., Sutherland, L., Cairney, J. & Booth, I.R. (1987). J. gen. Microbiol. **133**, 305.

Hua, S-S.T., Tsai, V.Y., Lichens, G.M. & Noma, A.T. (1982). Sci. Appl. Environ. Microbiol. **44**, 135.

Jakobsen, M. (1985). In Properties of Water in Foods (Simatos, D. & Multon, J.L., eds), p. 259. Martinus Nijhoff, Derdrecht, The Netherlands.

Kauzmann, W. (1948). Chem. Rev. **43**, 219.

Kennedy, E.P. (1982). Proc. Natl. Acad. Sci. USA **79**, 1092.

Kennedy, E.P. (1984). In The Cell Membrane: Its Role in Interaction with the Outside World (Haben, E., ed.), p. 33. Plenum Press, New York.

Kroll, R.G. & Anagnostopoulos, G.D. (1981). J. appl. Bact. **51**, 313.

Landfald, B. & Strom, A.R. (1986). J. Bact. **165**, 849.

Le Rudelier, D., Strom, A.R., Dandekar, A.M., Smith, L.T. & Valentine, R.C. (1984). Science, N.Y. **224**, 1064.

Lewis, J.C., Snell, N.S. & Burr, H.K. (1960). Science, N.Y. **132**, 544.

Mackenzie, A.P. (1977). Phil. Trans. Roy. Soc. **B. 278**, 167.

Martinek, K., Klibanov, A.M., Goldmacher, V.S., Tschernysheva, A.V., Moshaez, V.V., Berezin, I.V. & Glotov, B.O. (1977). Biochim.Biophys. Acta **485**, 13.

Milner, J.L., McClellan, D.J. & Wood, J.M. (1987). J. gen. Microbiol. **133**, 1851.

Mossel, D.A.A. & Westerdijk, K. (1949). Leeuwenhoek Med. Tidschr. **15**, 190.

Prior, B.A., Kenyon, C.P., van der Veen, M. & Mildenhall, J.P. (1987). J. appl. Bact. **62**, 119.

Russell, N.J., Kogut, M. & Kates, M. (1985). J. gen. Microbiol. **131**, 781.

Scott, W.J. (1953). Aust. J. Biol. Sci. **6**, 549.

Scott, W.J. (1957). Adv. Food Res. **7**, 84.

Slade, L. & Levine, H. (1985). Abstract; In Water Activity: A Credible Measure of Technological Performance and Physiological Viability? Royal Society of Chemistry, Cambridge.

Sutherland, L., Cairney, J., Elmore, M.J., Booth, I.R. & Higgins, C.F. (1986). J. Bacteriol. **168**, 805.

Tisa, L.S., Koshikawa, T. & Gerhardt, P. (1982). Appl. Environ. Microbiol. **43**, 1307.

Troller, J.A. & Christian, J.H.B. (1978). Water Activity and Food. Academic Press, New York.

Vaamonde, G., Scarmato, G., Chirife, J. & Parada, J.L. (1986). Lebensm. - Wiss. u-Technol. **19**, 403.

Warth, A.D. (1985). In Fundamental and Applied Aspects of Bacterial Spores (Dring, G.J., Ellar, D.J. & Gould, G.W., eds), p. 209, Academic Press, London.

MOULDS AND YEASTS ASSOCIATED WITH FOODS OF REDUCED WATER ACTIVITY: ECOLOGICAL INTERACTIONS

AILSA D. HOCKING
CSIRO Division of Food Research
P.O. Box 52, North Ryde, NSW 2113, Australia

INTRODUCTION

Food is a rich habitat for microorganisms, containing an abundance of nutrients such as carbohydrate, proteins and lipids, and growth factors such as vitamins and minerals. Food preservation and processing systems aim to prevent or inhibit microbial spoilage of foods, without diminishing their nutrient value or their acceptability.

Preservation of foods by reducing their water activity (a_w) is indeed an ancient technology. It may be achieved directly by dehydration, such as sun-drying or oven-drying, or by the addition of solutes such as sugar and/or salt. Often, a combination of these methods is used. In many foods, a_w is the dominant factor governing food stability, and it has a strong influence over the types of microorganisms able to grow in, and spoil, foods (Table 1). At high a_w (above 0.95), bacteria are the dominant flora of most foods. At lower a_w values, yeasts and moulds take over as the major spoilage organisms. Below about 0.85 a_w, foods are rarely spoiled by bacteria. The exceptions are spoilage of brines and salted foods by moderately and extremely halophilic bacteria. Bacteria do not appear to have adapted to sugar-rich environments. Below 0.85 a_w, the filamentous fungi are by far the most numerous and most diverse group of microorganisms.

The interactions between a_w and pH are also extremely important in selecting the dominant microflora of a particular food (Fig. 1). At neutral pH and high a_w (above 0.95), bacteria are the most important, but in low pH foods (below about 4.0), yeasts are likely to be dominant, although many _Lactobacillus_ species grow well, and some filamentous fungi also compete quite well in high a_w, low pH environments.

In the a_w ranges of intermediate moisture foods (IMF), i.e. about 0.90-0.75 a_w, yeasts and moulds are the most significant spoilage organisms, although halophilic bacteria may be a problem in some salty foods. Different types of IMF have characteristic mycofloras associated with them. The factors that determine which types of fungi are most common on particular foods are a_w, type of solute (sugar or salt) and storage temperature and conditions. Other factors such as the presence of food additives (spices, flavours, preservatives) and the carbon:

TABLE 1
Water activity and microbial water relations in perspective

a_w	Perspective	Foods	Moulds	Yeasts
1.00	Blood Plant wilt point Seawater	Vegetables Meat, Milk Fruit		
0.95	Most bacteria	Bread	Basidio- mycetes Most soil fungi	Basidio- mycetes
		Ham	Mucorales	
0.90			*Fusarium* *Cladosporium*	Most ascomycetes
	Staphylococcus		*Rhizopus*	
0.85		Dry salami	*Aspergillus* *flavus*	*Saccharomyces* *rouxii* (salt)
0.80			Xerophilic Penicillia	*Saccharomyces* *bailii*
0.75	Salt lake Halophiles	Jams Salt fish Fruit cake Confec- tionery	Xerophilic Aspergilli *Wallemia*	*Debaryomyces*
0.70		Dried fruit	*Eurotium* *Chrysosporium*	
0.65				
			Xeromyces *bisporus*	*Saccharomyces* *rouxii* (sugar)
0.60	DNA disordered			

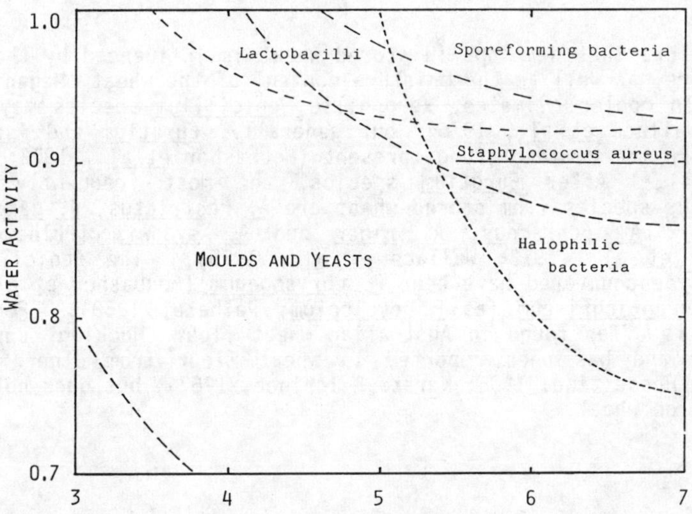

Figure 1. A schematic diagram of the combined influence of pH and a_w on microbial growth.

nitrogen balance of the food (i.e. whether it is a high carbohydrate or a high protein food) influence the composition of the mycoflora.

Intermediate moisture foods can be divided into four broad groups, for the purpose of examining the characteristic mycofloras: (1) cereals, nuts and spices; (2) dried fruits and confectionery; (3) dried meats; and (4) dried seafoods. Each of these groups is dealt with below.

CEREALS, NUTS AND SPICES

The mycoflora of grains and nuts changes during storage. Immediately after harvest, the dominant fungi on grains are field fungi, such as Alternaria, Fusarium, Cladosporium, Penicillium, and yeasts and smuts (Christensen & Kaufmann, 1965). However, these fungi do not have the capacity to grow or cause spoilage unless the a_w is above about 0.90. During storage, the field fungi gradually die out and storage fungi take their place. The most important genera are Eurotium (also known as the Aspergillus glaucus group), Aspergillus, particularly members of the A. restrictus series, some of the more xerophilic Penicillium species, Wallemia and occasionally Chrysosporium (Christensen & Kaufmann, 1965; Christensen, 1978; Sauer et al., 1984; Pitt & Hocking, 1985a).

Wheat

The mycoflora that develops in stored wheat is influenced by the storage temperature, as well as the moisture content of the wheat (Magan & Lacey, 1985). In cooler climates, xerophilic Penicillium species may be most common (Wallace et al., 1976), but generally, Eurotium and Aspergillus species are the dominant fungi present (Moubasher et al., 1972; Sauer et al., 1984). After Eurotium species, the most frequently reported Aspergillus species from stored wheat are A. restrictus, A. candidus, A. versicolor, A. ochraceus, A. niger and A. sydowii (Pelhate, 1968; Moubasher et al., 1972; Wallace et al., 1976). The Penicillia most frequently encountered have been P. chrysogenum (Moubasher et al., 1972) and P. aurantiogriseum (as P. cyclopium; Pelhate, 1968). Penicillium citrinum is often found in Australian wheat flour (Hocking, unpublished results), and has been reported in wheat flour from other countries (Graves & Hesseltine, 1966; Kurata & Ichinoe, 1967), but does not seem to be common on wheat.

Rice

The mycoflora of rice reflects the fact that it is a warmer climate crop than most other grains. Aspergillus and Eurotium species are again predominant, with E. repens, E. rubrum, E. amstelodami and E. chevalieri the most common Eurotium species, together with A. restrictus, A. candidus, A. versicolor and A. ochraceus. Of the Penicillium species identified, P. rugulosum, P. citreonigrum (as P. citreoviride), P. canescens, P. aurantiogriseum (as P. cyclopium) and P. citrinum have been recorded most commonly (Kurata et al., 1968; Mallick & Nandi, 1981).

Barley

The mycoflora of stored barley is similar to that of stored wheat. Common species reported in an Egyptian study were Aspergillus fumigatus, A. niger, A. flavus, A. sydowii, Penicillium citrinum and P. funiculosum (Abdel-Kader et al., 1979). From the cooler climate of Scotland, Flannigan (1969) reported that Eurotium species were most frequently isolated, followed by Penicillium species, and a few species of Aspergillus.

Maize

Because maize requires a high humidity for growth, persistent moist conditions during harvest and drying may allow some field fungi to become well established. Preharvest invasion by toxigenic species such as Aspergillus flavus, Fusarium graminearum and F. moniliforme may lead to production of aflatoxins and trichothecene or other Fusarium toxins. However, these fungi are not normally the dominant flora of stored maize. In studies of the mycoflora of U.S. maize, Lichtwardt et al. (1958) and Barron and Lichtwardt (1959) showed that Eurotium species, especially E. rubrum, E. amstelodami and E. chevalieri were the most significant spoilage fungi, together with Aspergillus restrictus and Penicillium species, especially P. aurantiogriseum and P. viridicatum.

Commodities that are made from grains, such as flour, semolina, and related high carbohydrate foods like sago, usually have a similar mycoflora to the grains from which they were derived.

Nuts

Dried nuts are susceptible to fungal spoilage because their soluble carbohydrate content is low, and their oil content is high. Consequently, any increase in moisture content has a significant effect on the a_w. Such a rise can easily be caused by moisture migration due to uneven temperatures in shipping containers, or direct sunlight on one side of a poorly insulated storage silo. Although many different species of fungi have been found in stored nuts, few of them have the capacity to cause spoilage unless the a_w rises to an unacceptably high figure. In stored peanuts, Aspergillus niger has been reported as a significant spoilage species (Joffe, 1969). Other species commonly isolated from stored peanuts are A. flavus, Penicillium funiculosum and P. purpurogenum (Joffe, 1969). These latter species are unlikely to cause spoilage unless the a_w is quite high. In an extensive study on the mycoflora of pecan nuts (Huang & Hanlin, 1975), 119 species from 44 genera were isolated from 37 samples of nuts. The most common species, in order of dominance, were Aspergillus niger, A. flavus, Eurotium repens, E. rubrum, A. parasiticus, A. ficuum, Rhizopus oryzae and Penicillium expansum. A study on hazelnuts (Senser, 1979) found that Rhizopus stolonifer and Penicillium aurantiogriseum were the most common fungi.

In our experience, the most common species causing spoilage of nuts and grains during storage and/or shipment are Eurotium species and Aspergillus restrictus or A. penicilloides. If commodities become extremely wet, then other Aspergillus species or Penicillia may cause spoilage.

Xerophilic fungi cause rancidity and other off-flavors in nuts, oilseeds and grains, as well as decreasing their germinability. If commodities become sufficiently wet (above about 0.85 a_w), mycotoxins may be formed by Aspergillus and Penicillium species.

Spices

Spices are frequently heavily contaminated with xerophilic fungi, with figures of up to 10^9 per gram being recorded (Hocking, 1981). The fungi most commonly isolated from spices are Eurotium species, Aspergillus restrictus and A. penicilloides, Wallemia sebi and other Aspergillus species. In our laboratory, we have isolated some of the more fastidious xerophiles, like Xeromyces bisporus and xerophilic Chrysosporium species from spices, but there are no reports in the literature of spice contamination by these fungi. Xerophilic fungi may cause spoilage of spices by decreasing the flavor compounds and volatile components, producing off-flavours, and causing ground spices to clump. Spices carrying high numbers of xerophilic fungal spores may contaminate the products in which they are used, possibly shortening the shelf-life of those products.

DRIED FRUITS AND CONFECTIONERY

Foods that contain high concentrations of sugar, e.g. dried fruits, confectionery, jams and conserves, jellies, fruit cakes, and fruit concentrates, are susceptible to spoilage by the same range of xerophilic fungi as the previous group of foods. However, they also provide an ideal habitat for some less common, more fastidious xerophilic fungi: Xeromyces bisporus, xerophilic Chrysosporium species, Eremascus species and the xerophilic yeast, Saccharomyces rouxii (also known as Zygosaccharomyces rouxii).

Jams

Jams and conserves are usually made to about 0.78 to 0.75 a_w. They are hot-filled and the jars vacuum-sealed. Spoilage rarely occurs in unopened jars. Spoilage may occur in opened jars, and the fungi most often responsible are Eurotium species and Saccharomyces rouxii. In jams of higher a_w, xerophilic Penicillia, such as P. corylophilum may cause spoilage (Pitt & Hocking, 1985a).

Dried Fruits

Many dried fruits, such as apricots, pears, peaches and bananas contain high levels of SO_2, which prevents browning, and eliminates the microflora. However, apricots with slightly increased moisture content (above 0.72 a_w) may be spoiled by the xerophilic yeasts Saccharomyces rouxii and S. bailii which can tolerate low levels of SO_2.

Some dried fruits are produced without SO_2, such as prunes and dried vine fruits. Pitt and Christian (1968) isolated a wide variety of xerophilic moulds from Australian dried and high moisture prunes. The fungi most commonly isolated were Eurotium species, especially E. herbariorum, Xeromyces bisporus and xerophilic Chrysosporium species. Prunes are the only substrate from which the rare fungi Eremascus albus and E. fertilis have been reported in recent years (Pitt & Christian, 1968). Tanaka and Miller (1963), in a survey of the mycoflora of dried prunes, found 13 species of yeasts, the most common being S. rouxii followed by S. mellis, Torulopsis magnoliae and T. stellata. They also isolated 124 strains of moulds, principally Eurotium species (45.2%), Aspergillus niger (14.5%) and various Penicillium species (33%).

Xeromyces bisporus has also been reported from dried prunes in the United Kingdom (Dallyn & Everton, 1969). This fungus appears to be quite common on dried fruit, having been found in currants, 0.66-0.67 a_w (Dallyn & Everton, 1969); Chinese dates, 0.72 a_w (Pitt & Hocking, 1982); and in Australian muscatel raisins (0.66 a_w) and dates from the Middle East (Hocking, unpublished). Wallemia sebi is rarely isolated from dried fruit, but has been reported spoiling dried paw-paw (papaya) (Dallyn & Fox, 1980).

Confectionery

There is little in the literature on spoilage of confectionery by fungi. Most confectionery is manufactured with an a_w sufficiently low to prevent the growth of common xerophiles like Eurotium species. However, soft fillings in chocolates may be spoiled by Saccharomyces rouxii, which produces gas, splitting the chocolate casing and allowing the filling to leak out (Pitt & Hocking, 1985a). Xerophilic Chrysosporium species have been isolated from table jellies, coconut (Kinderlerer, 1984b), and from jelly confections (Pitt & Hocking, 1985a). Coconut can be spoiled by a number of xerophilic moulds. Eurotium species and P. citrinum cause ketonic rancidity, while growth of Chrysosporium species causes cheesy butyric off-flavors (Kinderlerer, 1984a). Xeromyces bisporus, the most xerophilic of all the fungi, has been isolated from licorice (Pitt & Hocking, 1982; Pitt & Hocking, 1985a), table jelly (Dallyn & Everton, 1969), chocolate sauce, 0.77 a_w (Dallyn & Everton, 1969), and a marshmallow chocolate filling (Hocking, unpublished).

Fruit Concentrates, Syrups, Honey and Brines

Low a_w liquid products (between 0.60 and 0.85 a_w) are susceptible to spoilage by xerophilic yeasts. In high sugar environments, the most common species are Saccharomyces rouxii, S. bisporus and S. bailii and some Torulopsis species (Tilbury, 1976; Pitt & Hocking, 1985a). In brines, Debaryomyces hansenii (Torulopsis famata) is probably the most significant spoilage yeast (Pitt & Hocking, 1985a).

DRIED AND MANUFACTURED MEAT PRODUCTS

Many manufactured meat products such as salamis, hams and other cured meats have a relatively high a_w of 0.80-0.95. These products rely on a combination of factors for their microbiological stability, e.g. nitrite, reduced pH, addition of salt, reduction of the Eh by vacuum packaging, and sometimes, a heat treatment during manufacture (Leistner, 1985). Dried meats, such as biltong and Chinese dried meat products (sou-gan) rely primarily on reduced a_w for their stability. Consequently, the mycofloras of these two classes of meat products are somewhat different.

Salamis, wursts and cured meats are most likely to be spoiled by Penicillium species, though Eurotium, Aspergillus and other moulds also occur (Leistner & Ayres, 1968; Takatori et al., 1975; Hadlok et al., 1976). A Japanese study on imported salamis (Takatori et al., 1975) found that the most common Penicillium species were P. cyclopium (=P. aurantiogriseum), P. miczynskii and P. viridicatum, with Aspergillus, Eurotium, Cephalosporium, Chrysosporium and Mucor species also reported. Extensive studies of the mycoflora of fermented sausages and cured hams in Europe (Hadlok et al., 1976) and U.S.A. (Leistner & Ayres, 1968) reported that the most prevalent mould genera on these products were Penicillium, Aspergillus, Eurotium and Cladosporium. Frequently isolated Penicillium species were P. expansum, P. janthinellum and P. chrysogenum in the U.S. study (Leistner & Ayres, 1968) and P. aurantiogriseum, P. palitans (P. solitum) and P. chrysogenum on European sausages (Hadlok et al., 1976). Both studies reported E. rubrum and E. repens as the most common Eurotium species, and A. versicolor the most common Aspergillus species. Other frequently isolated Aspergilli were A. fumigatus and A. tamarii (Hadlok et al., 1976), and A. restrictus, A. niger and A. wentii (Leistner & Ayres, 1968). Hadlok (1969) noted that spices play a major role in the contamination of meat products with Penicillium and Aspergillus.

Dried meats have received less attention in the microbiological literature, no doubt because they are intrinsically more stable. African biltong is a dried meat product which is brined or dry salted. Pickling mixtures may also include brown sugar, vinegar, pepper, coriander and other spices. The meat is air-dried for 1-2 weeks after curing (van der Riet, 1982). Leistner (1985) reported that biltong was stable if it was less than 0.77 a_w and less than pH 5.5, but van der Riet (1982) suggested that biltong should be 0.68 a_w (24% moisture) or less to prevent mould growth. In a study of the mycoflora of 20 commercial biltong samples (van der Riet, 1976), Eurotium, Aspergillus and Penicillium were the dominant genera, with the most common spoilage species being E. amstelodami, E. chevalieri, E. repens, E. rubrum, A. versicolor and A. sydowii. The most frequently isolated yeasts were Debaryomyces hansenii, Candida zeylanoides and Trichosporon cutaneum (van der Riet, 1976). Leistner (1985) reported that Chinese intermediate moisture meat products could be spoiled by Eurotium species if the a_w was greater than 0.69.

DRIED SEAFOOD PRODUCTS

Fish and other seafoods constitute a major source of protein in the diet of many people in tropical countries. Salting and drying are the most widespread and cheapest methods of fish preservation, but the a_w of the products is often not low enough to prevent mould spoilage. As with dried meat products, the fungi most commonly reported from dried seafoods are Eurotium species, followed by Aspergillus and Penicillium (Okafor, 1968; Townsend et al., 1971; Hitokoto et al., 1976; Ichinoe et al., 1977; Phillips & Wallbridge, 1977; Ogasawara et al., 1978; Wu & Salunkhe, 1978; Wheeler et al., 1986). Scopulariopsis, Cladosporium, Wallemia, Mucor and Acremonium species have also been reported, but in lower numbers.

The most frequently isolated Eurotium species are E. rubrum, E. repens, E. amstelodami and E. chevalieri. The most common Aspergillus species are A. restrictus, A. niger, A. versicolor series, and A. flavus. Aspergillus wentii was reported from nearly 15% of samples of dried fish from Indonesia (Wheeler et al., 1986), but has rarely been reported elsewhere. Penicillium species appear to play only a minor role in spoilage of dried seafoods in tropical areas, although they are isolated often. Penicillia are more important in temperate climates. In a study of the mycoflora of dried shrimps in the U.S.A. (Wu & Salunkhe, 1978), almost half the fungi isolated were Penicillium species, and Penicillia were reported as comprising over 20% of the mycoflora of dried sardines in a Tokyo market (Hitokoto et al., 1976).

In a recent study on the mycoflora of Indonesian dried fish and seafoods (Wheeler et al., 1986), the principal fungus isolated was a previously undescribed species of white mould, Polypaecilum pisce (Pitt & Hocking, 1985b). It was isolated from nearly 40% of the 74 samples examined, in some cases covering the fish in a powdery, white growth. In view of its prevalence on these samples, it is remarkable that this species has not been described before. The halophilic fungus <u>Basipetospora</u> <u>halophila</u> (<u>Scopulariopsis</u> <u>halophilica</u>) (Pitt & Hocking, 1985b) was also isolated during the survey of moulds on Indonesian dried seafoods, though it was less common, being isolated from only 5 samples (Wheeler <u>et</u> <u>al</u>., 1986). <u>B.</u> <u>halophilica</u> is a relatively rare fungus, and all reported isolates have come from dried fish or seaweed (Pitt & Hocking, 1985a).

<u>Wallemia</u> <u>sebi</u> was once regarded as the principal fungus spoiling dried and salted fish (Frank & Hess, 1941), on which it is known as "dun" mould. However, it is rare on tropical fish; most reports of <u>Wallemia</u> on dried fish products have come from temperate regions (Frank & Hess, 1941; Hitokoto <u>et</u> <u>al</u>., 1976).

FACTORS AFFECTING THE MYCOFLORA OF INTERMEDIATE MOISTURE FOODS

Water activity is no doubt the principal factor that selects for a particular fungal flora on a particular foodstuff. Many other factors also exert an important selective pressure, and among the most important are solute type (i.e. sugar or salt) and storage temperature. The competitiveness of a particular species in a particular environment relies very much on its growth rate, and its ability to reproduce rapidly. There have been numerous studies on the effects of a_w, solute and temperature on the growth rates of spoilage fungi (Ayerst, 1969; Hocking & Pitt, 1979; Avari & Allsopp, 1983; Andrews & Pitt, 1987; Wheeler <u>et</u> <u>al</u>., submitted; Wheeler <u>et</u> <u>al</u>., in preparation, a & b). It is obvious from these studies that some fungi compete much better in salty environments, and some are better adapted for high sugar environments.

Figs 2 and 3 show the growth rates of five fungi on salt-based media at 20°C and 30°C, respectively. Two of these fungi, <u>Polypaecilum</u> <u>pisce</u> and <u>Basipetospora</u> <u>halophila</u>, are halophilic (Andrews & Pitt, 1987; Wheeler <u>et</u> <u>al</u>., in preparation, b). <u>Wallemia</u> <u>sebi</u> was once considered to be halophilic because it was isolated from salted fish in the Northern hemisphere (Frank & Hess, 1941), but it is now known that it actually grows faster in sugary environments (Pitt & Hocking, 1977, Wheeler, <u>et</u> <u>al</u>., submitted).

The effects of a_w and salt on the comparative growth rates at 20°C of five fungi from dried fish are shown in Fig. 2. At high a_w in a salty environment, <u>A.</u> <u>flavus</u> would outgrow the other four fungi illustrated down to about 0.925 a_w. Between 0.925 and 0.89 a_w, <u>E.</u> <u>rubrum</u> has the fastest growth rate. Below 0.89 a_w, the three halophilic fungi grow better than the <u>Aspergillus</u> or <u>Eurotium</u>. Although <u>Wallemia</u> <u>sebi</u> has a very slow growth rate, it competes very well because it produces huge numbers of very small conidia, and it produces them rapidly, often within 48 hours of germination (Hocking, 1985). <u>Polypaecilum</u> <u>pisce</u> competes relatively poorly at 20°C, growing only down to 0.86 a_w, while <u>Basipetospora</u> <u>halophila</u> and <u>W.</u> <u>sebi</u> both grow down to 0.75 a_w, i.e. saturated NaCl, at this temperature.

Fig. 3 shows growth rates of the same five fungi on salt-based media at 30°C. <u>Aspergillus</u> <u>flavus</u> again dominates at the higher a_w values, down to 0.92-0.91 a_w, with <u>Eurotium</u> <u>rubrum</u> competing well around 0.93-0.90 a_w. The species that obviously competes best from about 0.90 down to 0.75 a_w, is <u>P.</u> <u>pisce</u>, with <u>B.</u> <u>halophila</u> also growing well, particularly below 0.80 a_w. It is clear that <u>W.</u> <u>sebi</u> does not compete well at 30°C, ceasing growth at 0.86 a_w. The growth data illustrated in Figs 2 and 3 support the observations that P. pisce is often the dominant fungus on dried salted fish from the tropics, while W. sebi is rarely isolated from such samples (Wheeler et al., 1986). The a_w values of fish sampled in the Indonesian study were between 0.79 and 0.65 a_w (Wheeler et al., 1986). During drying, the a_w of the fish would be between 0.90 and 0.75 for some time, ideal conditions for growth of fungi like P. pisce and B. halophila. P. pisce can grow down to 0.705 a_w on glucose-fructose based media at 30°C (Wheeler et al., in preparation, b), and the growth curve in Fig. 3 implies that this species would be able to grow at less than 0.75 a_w in salty conditions as well.

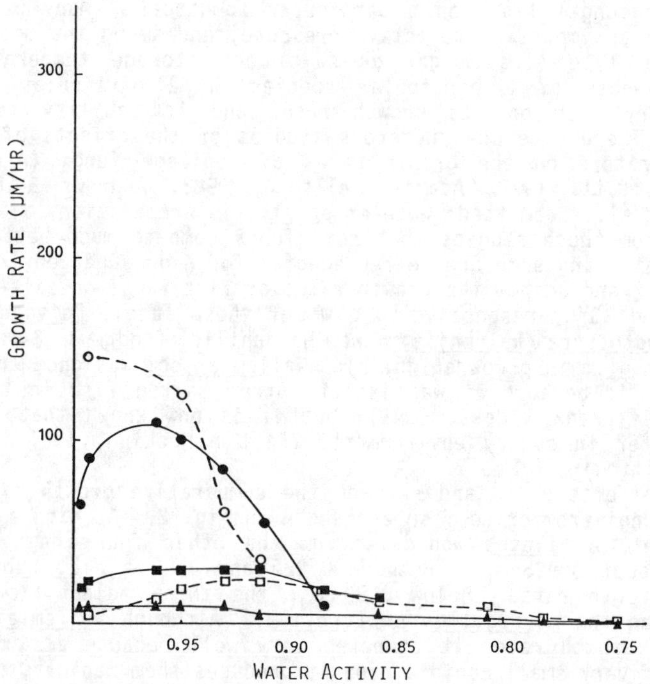

Figure 2. Comparative growth rates of five fungi on NaCl-based media at 20°C. (●) Eurotium rubrum; (○) Aspergillus flavus; (■) Polypaecilum pisce; (□) Basipetospora halophila; (▲) Wallemia sebi.

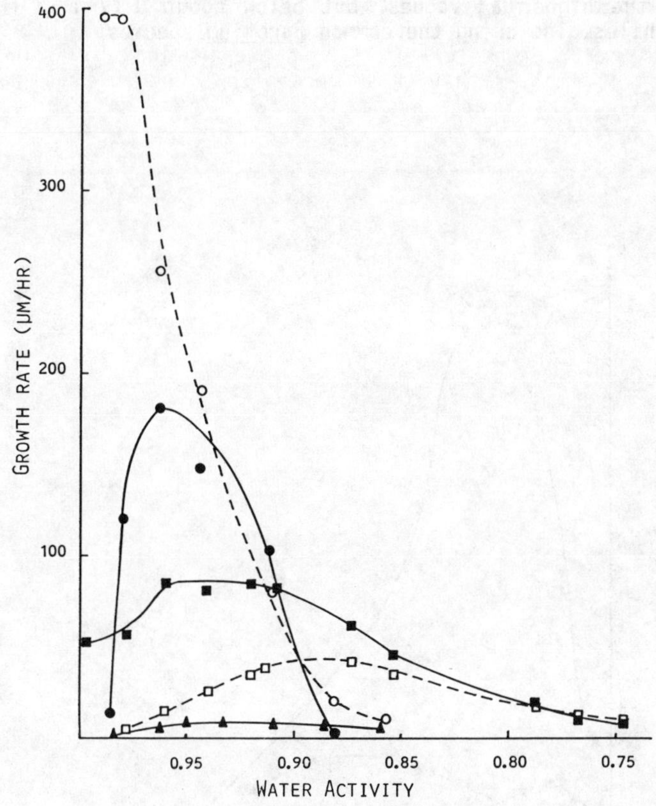

Figure 3. Comparative growth rates of five fungi on NaCl-based media at 30°C. (●) Eurotium rubrum; (○) Aspergillus flavus; (■) Polypaecilum pisce; (□) Basipetospora halophila; (▲) Wallemia sebi.

Substrates with a high sugar content have a different mycoflora from salty foods. Fig. 4 shows the comparative growth rates of five fungi on glucose-fructose based media at 25°C: Eurotium rubrum, E. halophilica, Polypaecilum pisce, Xeromyces bisporus and Chrysosporium fastidium. E. rubrum is representative of most Eurotium species. It grows extremely rapidly on sugary substrates, with an optimum between about 0.98 and 0.85 a_w. Below about 0.77 a_w, its growth rate is less than that of several of the fastidious extreme xerophiles. P. pisce competes very poorly in sugary environments at 25°C and would probably be overgrown by any Eurotium species present. The three extreme, obligate xerophiles illustrated, X. bisporus, E. halophilicum and C. fastidium, compete poorly at the higher a_w values, but below about 0.77 a_w, will outgrow most xerophiles, including the common Eurotium species.

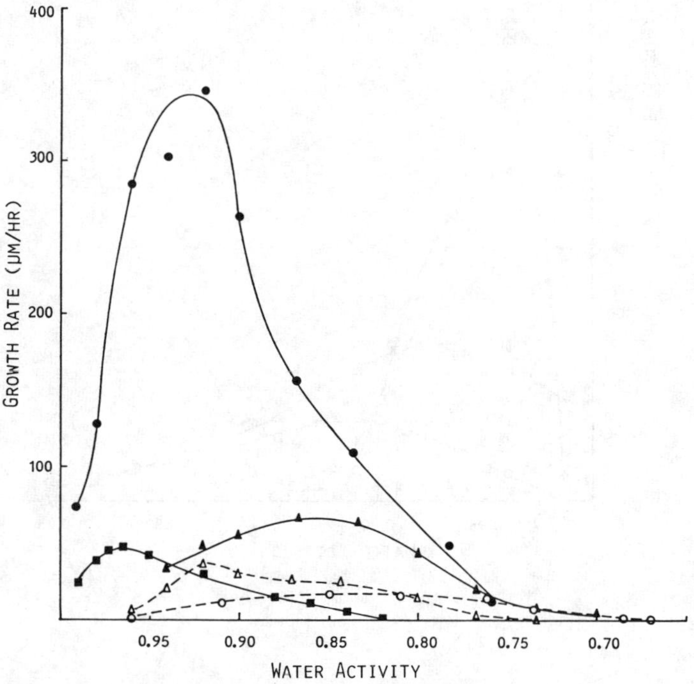

Figure 4. Comparative growth rates of five fungi on glucose/fructose based media at 25°C. (●) Eurotium rubrum; (○) E. halophilicum; (■) Polypaecilum pisce; (▲) Xeromyces bisporus; (△) Chrysosporium fastidium.

Eurotium halophilicum is an exceptionally and obligately xerophilic Eurotium; its water relations were studied by Andrews and Pitt (1987), as Eurotium species FRR 2471. This isolate was subsequently identified as E. halophilicum (Hocking & Pitt, submitted). E. halophilicum is unable to grow above 0.94 a_w, and has a very slow growth rate, even under optimum conditions. It is also misnamed, in that it is xerophilic rather than halophilic, growing faster and at a lower a_w in glucose-fructose based media than in salt-based media (Andrews & Pitt, 1987).

Xeromyces bisporus is the most xerophilic of all known fungi. It has been reported as germinating at 0.605 a_w, and able to produce ascospores at 0.67 a_w and aleurioconidia at 0.66 a_w (Pitt & Christian, 1968). In my experience, X. bisporus is probably also the most common of the extremely xerophilic fungi.

No doubt, other factors are important in determining which fungi are most likely to colonise a particular habitat or foodstuff. The carbon:nitrogen ratio, the pH, the presence of preservatives, the type of packaging and the gaseous atmosphere will all influence the mycoflora. However, few studies on the importance of these factors, or the interactions between them, have been carried out.

PREVENTION OF SPOILAGE OF INTERMEDIATE MOISTURE FOODS BY XEROPHILIC FUNGI

If the a_w of an IMF cannot be reduced sufficiently to prevent fungal growth, then additional inhibitory hurdles can be used (Leistner, 1985). Heat processing will kill most fungal spores, so heating at some stage during the production of an IMF will significantly lower the load of fungal spores. However, ascospores are more resistant to heat than conidia, e.g. most Aspergillus conidia are killed by a heat treatment of 60°C for 10 minutes, but some ascospores, particularly those of E. chevalieri, can survive 80°C for 10 minutes (Pitt & Christian, 1970). The ascospores of Xeromyces bisporus are also quite heat-resistant. Pitt & Christian (1970) reported that a small proportion (0.1%) survived 10 minutes at 80°C, while Dallyn & Everton (1969) observed that 2 minutes at 90°, 4 minutes at 85° or 9 minutes at 80°C was required to kill 2000 X. bisporus spores in a medium of 0.9 a_w and pH 5.4. X. bisporus has caused spoilage of Australian fruit cakes, 0.75-0.76 a_w, because the ascospores survived the baking process (Pitt & Hocking, 1982).

Preservatives can often prevent or delay mould spoilage of IMF, and potassium sorbate or benzoate are now added to some types of IMF to prevent spoilage, e.g. high moisture prunes, some IM meat products and some baked goods (Bolin & Boyle, 1967; Schade *et al*., 1973; Leistner, 1985). A heat process in the presence of preservatives is more effective in killing fungal conidia than heat alone (Beuchat, 1981), so the combination is an even more efficient way of prolonging the shelf-life of IMF. New trends in vacuum packaging and modified atmosphere storage are also helping to extend the shelf-life of many IMF by delaying both rancidity and fungal spoilage.

ACKNOWLEDGEMENT

Thanks are expressed to Ms Kathy Wheeler for allowing use of some of her water activity data before publication.

REFERENCES

Abdel-Kader, M.I.A., Moubasher, A.H. & Abdel-Hafez, S.I.I. (1979). Mycopathologia **68**, 143.

Andrews, S. & Pitt, J.I. (1987). J. gen. Microbiol. **133**, 233.

Avari, G.P. & Allsopp, D. (1983). Biodeterioration **5**, 548.

Ayerst, G. (1969). J. Stored Prod. Res. **5**, 669.

Barron, G.L. & Lichtwardt, R.W. (1959). Iowa St. J. Sci. **34**, 147.

Beuchat, L.R. (1981). Appl. environ. Microbiol. **41**, 472.

Bolin, H.R. & Boyle, F.P. (1967). J. Sci. Food Agric. **18**, 289.

Christensen, C.M. (1978). In Food and Beverage Mycology (Beuchat, L.R., ed.), p. 173. Avi Publishing Company, Westport, Connecticut.

Christensen, C.M. & Kaufmann, H.H. (1965). Annu. Rev. Phytopath. **3**, 69.

Dallyn, H. & Everton, J.R. (1969). J. Food Technol. **4**, 399.

Dallyn, H. & Fox, A. (1980). In Microbial Growth and Survival in Extremes of Environment (Gould, G.W. & Corry, J.E.L., eds), p. 129. Academic Press, London.

Flannigan, B. (1969). Trans. Br. mycol. Soc. **53**, 371.

Frank, M. & Hess, E. (1941). J. Fish. Res. Bd Can. **5**, 276.

Graves, R.R. & Hesseltine, C.W. (1966). Mycopath. Mycol. appl. **29**, 277.

Hadlok, R. (1969). Fleischwirtschaft **49**, 1601.

Hadlok, R., Samson, R.A., Stolk, A.C. & Schipper, M.A.A. (1976). Fleischwirtschaft **56**, 374.

Hitokoto, H., Morozumi, S., Wauke, T., Sakai, S., Zen-Yoji, H. & Benoki, M. (1976). Annu. Rep. Tokyo Metrop. Res. Lab. Public Hlth **27**, 36.

Hocking, A.D. (1981). CSIRO Food Res. Q. **44**, 73.

Hocking, A.D. (1985). J. gen. Microbiol. **132**, 269.

Hocking, A.D. & Pitt, J.I. (1979). Trans. Br. mycol. Soc. **73**, 141.

Hocking, A.D. & Pitt, J.I. Two new xerophilic fungi and a further record of Eurotium halophilicum. Mycologia (submitted).

Huang, L.H. & Hanlin, R.T. (1975). Mycologia **67**, 689.

Ichinoe, M., Suzuki, M. & Kurata, H. (1977). Bull. Nat. Inst. hyg. Sci. **95**, 96.

Joffe, A.Z. (1969). Mycopath. Mycol. appl. **39**, 255.

Kinderlerer, J.L. (1984a). Food Microbiol. **1**, 23.

Kinderlerer, J.L. (1984b). Food Microbiol. **1**, 205.

Kurata, H. & Ichinoe, M. (1967). J. Food hyg. Soc. Japan **8**, 237.

Kurata, H., Udagawa, S., Ichinoe, M., Kawasaki, Y., Takada, M., Tazawa, M., Koizumi, A. & Tanabe, H. (1968). J. Food hyg. Soc. Japan **9**, 23.

Leistner, L. (1985). In Properties of Water in Foods (Simatos, D. & Multon, J.L., eds), p. 309. Martinus Nijhoff, Dordrecht.

Leistner, L. & Ayres, J.C. (1968). Fleischwirtschaft **48**, 62.

Lichtwardt, R.W., Barron, G.L. & Tiffany, L.H. (1958). Iowa St. Coll. J. Sci **33**, 1.

Magan, N. & Lacey, J. (1985). Trans. Br. mycol. Soc. **85**, 29.

Mallick, A.K. & Nandi, B. (1981). Rice J. **84**, 8.

Moubasher, A.H., Elnaghy, M.A. & Abdel-Hafez, S.I. (1972). Mycopath. Mycol. appl. **47**, 261.

Ogasawara, K., Sekijo, I., Sunagawa, H. & Umemura, M. (1978). Rep. Hokkaido Inst. Public Hlth **28**, 26.

Okafor, N. (1968). Niger. J. Sci. **2**, 41.

Pelhate, J. (1968). Bull. trimest. Soc. Mycol. Fr.l **84**, 127.

Phillips, S. & Wallbridge, A. (1977). In Proceedings of the Conference on the Handling, Processing and Marketing of Tropical Fish., p. 353. Tropical Products Institute, London, U.K.

Pitt, J.I. & Christian, J.H.B. (1968). Appl. Microbiol. **16**, 1853.

Pitt, J.I. & Christian, J.H.B. (1970). Appl. Microbiol. **20**, 682.

Pitt, J.I. & Hocking, A.D. (1977). J. gen. Microbiol. **101**, 35.

Pitt, J.I. & Hocking, A.D. (1982). CSIRO Food Res. Q. **42**, 1.

Pitt, J.I. & Hocking, A.D. (1985a). Fungi and Food Spoilage. Academic Press, Sydney.

Pitt, J.I. & Hocking, A.D. (1985b). Mycotaxon **22**, 179.

Sauer, D.B., Storey, C.L. & Walker, D.E. (1984). Phytopathology **74**, 1050.

Schade, J.E., Stafford, A.E. & King, A.D. (1973). J. Sci. Food Agric. **24**, 905.

Senser, F. (1979). Gordian **79**, 117.

Takatori, K., Takahashi, K., Suzuki, T., Udagawa, S. & Kurata, H. (1975). J. Food hyg. Soc. Japan **16**, 307.

Tanaka, H. & Miller, M.W. (1963). Hilgardia **34**, 167.

Tilbury, R.H. (1976). In Intermediate Moisture Foods (Davis, R., Birch, C.G., & Parker, K.J., eds), p. 138. Applied Science Publ., London.

Townsend, J.F., Cox, J.K.B., Sprouse, R.F. & Lucas, F.V. (1971). J. Trop. Med. Hyg. **74**, 98.

Van der Riet, W.B. (1976). S. Afr. Food Rev. **3(1)**, 105.

Van der Riet, W.B. (1982). Fleischwirtschaft **62**, 1000.

Wallace, H.A.H., Sinha, R.N. & Mills, J.T. (1976). Can. J. Bot. **54**, 1332.

Wheeler, K.A., Hocking, A.D., Pitt, J.I & Anggawati, A. (1986). Food Microbiol. **3**, 351.

Wheeler, K.A., Hocking, A.D. & Pitt, J.I. Effects of temperature and water activity on germination and growth of Wallemia sebi. Trans. Br. mycol. Soc. (submitted).

Wheeler, K.A., Hocking, A.D. & Pitt, J.I. Water relations of some Aspergillus species isolated from dried fish. (a: in preparation for Trans. Br. mycol. Soc.)

Wheeler, K.A., Hocking, A.D. & Pitt, J.I. Influence of temperature on the water relations of Polypaecilum pisce and Basipetospora halophila, two halophilic fungi. (b: in preparation for J. gen. Microbiol.)

Wu, M.T. & Salunkhe, D.K. (1978). J. appl. Bacteriol. **45**, 231.

PRESERVATION OF SKIM-MILK POWDERS : ROLE OF WATER ACTIVITY AND TEMPERATURE IN LACTOSE CRYSTALLIZATION AND LYSINE LOSS

G. VUATAZ
Nestle Research Centre
Nestec Ltd., Vers-chez-les-Blanc
CH-1000 Lausanne 26, Switzerland

INTRODUCTION

Many physical and chemical reactions may occur in dehydrated foods either during drying or subsequently during storage with possible consequences for their preservation. Skim-milk powders offer a good example where both physical and chemical reactions are possible depending on the composition, the temperature, the water activity or other parameters. In this paper, we will focus our attention on lactose crystallization and lysine loss reactions. A review of the pertinent literature will first be given, followed by some personal experiences and results pointing out the role of water. Skim-milk powder is manufactured by spray-drying concentrated skimmed milk. It is consequently a widely used source of proteins and lactose. Its mean composition is the following:

- lactose 51%
- protein 36% (including 3% lysine)
- minerals 9%
- water 3%
- fat 1%

The only constituents we will consider in this paper are lactose, lysine and water. Our attention will be given to the chemical reaction between lactose and lysine, to the physical state of lactose and to the role of water on reactivity.

From the chemical aspect, the Maillard reaction involving lysine in different consecutive and competitive reactions may occur both during processing and storage, provoking the loss of available lysine. These chemical modifications of the milk proteins could have nutritional, metabolic and physiological consequences (Finot, 1983; Hurrell, 1984). For our purpose it is sufficient to remember that the "early" Maillard reaction (1) in milk powders is a reaction between lactose and lysine giving a relatively stable compound called lactose-lysine (or Amadori compound) which is not biologically available and should be avoided.

Only 0.04% water is produced by this reaction for 10% lysine loss.

lactose + lysine ———>lactose-lysine + H_2O (1)

From the physical aspect, the lactose, which is normally in the unstable amorphous state after rapid dehydration by spray-drying, may crystallize through an irreversible phase transition.

amorphous lactose ———>crystallized lactose (2)

The partial crystallization may appear in different proportions of the two anomeric forms: the alpha-lactose hydrate and the beta-lactose anhydride, depending on the available water and the temperature.

The rates of the reactions (1) and (2) which could occur in powders are influenced by the composition, temperature and water content. Technologists need a realistic description of the principal parameters involved in these reactions in order to avoid deterioration during drying or storage and to improve the quality of milk powders. It can easily be calculated from the above composition that, for one mole of lysine, milk products contain approximately 7 moles of lactose and skim-milk powder contains 8 moles of water, but the availability of the lactose depends on its physical state and the availability of the water depends on its binding state. The temperature is easy to measure and to control during experiments. Its effect on reactivity is generally well described by the Arrhenius equation which leads to calculation of the activation energy of the reaction. The water effects are more difficult to measure and to describe. First, the moisture distribution needs a long time to reach equilibrium and homogeneity in a sample due to mass transfer which is quite slow in the powders. Second, water can be bound in various ways with other molecules so that one part could be inert whereas another part could be free to participate in the reaction as a solvent. For these reasons it could be difficult to precisely determine the water content by extraction of the water using a reference method such as the oven method.

In fact, water content is not a sufficient parameter to describe the interactions of water with the other constituents. At present, it is better to consider the water activity (a_w) which characterises precisely the state of water and its availability for physical and chemical changes. The relation between the water activity and the water content is described by the empirical sorption isotherm (see Fig. 1) which is a characteristic of each dehydrated food. In particular, the sorption isotherm of amorphous and crystallized products are different. It must be noted that, for amorphous skim-milk powders, a 3% water content induces a low water activity of about 0.15. A phase transition will be represented by a step (breakpoint) from one curve to the other. The effect of water activity in relation with the water content will be discussed below for both the reactions of lysine loss and of lactose crystallization. Further, it will be described how lactose crystallization induces an increase of the water activity which may influence the Maillard reaction.

LITERATURE REVIEW

As our interest concerns both chemical and physical aspects, we have to consider a very abundant literature and the numerous parameters which are treated. An overview has been given by Labuza (1975) in the Proceedings of the first International Symposium on Water Relations of Foods. We will limit our review to the papers presenting results either on lactose crystallization in relation with sorption isotherms and moisture uptake, or on lysine loss due to the Maillard reaction in relation with water activity.

Sorption Isotherm and Lactose Crystallization

An extended review on structure transitions including collapse (structure sinking) and crystallization (phase transition) in freeze-dried and spray-dried sugars was given by Flink (1983). He pointed out that the irreversible transition from the metastable amorphous state to the stable crystalline state will have a large effect on the properties and quality of food products, especially of milk products containing lactose. Many authors have offered observations and explanations concerning moisture content determination, moisture uptake due to increase in the relative humidity, the break in the sorption isotherms of products containing amorphous sugars, temperature effects, various methods to analyse the structure of lactose in milk powders and the effects of lactose crystallization on the reconstitution properties of the powders.

In 1926, Supplee gave data showing the moisture sorption of various milk powders held at a relative humidity from 10 to 80% at 25°C. He observed the break in the sorption curve at approximately the same level of 50% (0.5 a_w) for all powders of the same protein content, irrespective of the method of manufacture and indicated that there is an area of instability between 40 and 50% (0.4 to 0.5 a_w). The desorption to the initial relative humidity does not give the same initial water content. The possibility of hydration of anhydrous lactose was considered to explain this hysteresis. This phenomenon has been clearly explained by King (1965): "due to high hygroscopicity of glassy lactose, dried milk particles very easily take up moisture whenever there is the opportunity. When this happens the concentration of lactose is diluted so that its molecules acquire sufficient mobility and space to arrange themselves into a crystal lattice."

The determination of the breakpoint (a_w and maximum water content) has been studied by many authors. Berlin _et al_. (1968) compared the water sorption isotherm due to protein and lactose components of milk powders. They found that at low a_w, water is bound primarily by the casein fraction. As the a_w progresses toward 0.5, lactose becomes more active in water binding and accumulates sufficient water to undergo transition from the amorphous to the crystalline hydrate form. They clearly demonstrated that lactose crystallization is the sole cause of the break usually observed in the water sorption isotherm for dehydrated dairy products. Linko _et al_. (1981) discussed the additive model of water adsorption by calculating the adsorption isotherm of skim-milk powder from its components. They obtained good agreement between adsorption isotherms of skim-milk powder and various binary and ternary mixtures. Casein appeared to be the main absorber of water at $a_w < 0.2$; in the range of 0.2 to 0.6 the sorption behaviour of dried milk products was dominated by the transformation of the physical state of lactose and at $a_w > 0.6$, salts present have a marked influence in water adsorption. Berlin _et al_. (1970) studied the effects of temperature on water sorption by milk powders. They noted that the breakpoint occurred at a lower a_w

when temperature was increased. Warburton & Pixton (1978) studied the effect of temperature on the water sorption by spray-dried skim-milk. The breakpoint was observed at 8.3% moisture when the temperature was raised from 25° to 35°C and at 7.4% between 35° and 45°C. The recommended limit of the water content to prevent crystallization which could lower the quality of the product was 6% at room temperature. Saltmarch & Labuza (1980) observed the break with release of water in the 0.33 to 0.44 a_w range for whey powder samples initially in the amorphous state. By scanning electron microscopy they showed that crystallization appears to cause rejection of other whey components from the lactose phase making them available to the environment.

Only a few authors give data concerning the kinetics of crystallization. Supplee (1926) noted that adsorption of water in skim-milk powder was rapid during the first few hours, reaching a maximum (the water content increased from 3% to 9.45% in 33 hours at 0.5 a_w and room temperature), and that there was a gradual decrease extending over a period of 8 weeks before constant weight was reached (6.38%). Henderson & Pixton (1980) discussed the kinetics of water adsorption by spray-dried skim-milk. From the change in weight, they measured the maximum and the final moisture contents in relation to the level of a_w assigned. At 0.55 a_w the maximum (11%) was reached after 45 hours and the final (8.2%) after 150 hours. At 0.72 a_w the maximum (13%) was reached after 20 hours and the final (12.3%) after 80 hours, but at 0.86 a_w no relative maximum was observed because the uptake of water by the protein was higher than the release of water by the crystallizing lactose. For sorption at 0.72 a_w, there was an unexplained difference between the final values calculated by weighing (12.3%) and the moisture content determined by oven drying (10.8%). This difference was not observed at 0.55 a_w. This could be due to the crystallization form which may change from the beta anhydride at 0.55 a_w to the alpha hydrate at 0.72 a_w (see Figs 7 to 9). Makower & Dye (1956) studied the rate of crystallization of amorphous sucrose, evaluated from changes in the water content. After an initial induction period, interpreted to be the time for build-up of sufficient nuclei to initiate an appreciable rate of crystallization, the rate of crystallization followed an exponential law with respect to time.

More recently, attention was given to the different crystallization forms which may occur depending on temperature, water activity and the initial beta/alpha ratio in the amorphous state of the lactose. X-ray diffraction, infrared spectroscopy and scanning electron microscopy are the techniques most often used both to test the crystallization state and distinguish the alpha or the beta form. King (1965) gave a review on the physical structure of dried milk including a chart describing the physico-chemical relationships between different forms of lactose. Usually the form which crystallizes out is alpha-lactose hydrate. However, under certain conditions of time and moisture content both the alpha hydrate and the beta anhydride may crystallize out together at room temperature. Bushill et al. (1965) made comparative studies of moisture sorption by spray- and freeze-dried milk and spray- and freeze-dried lactose. They showed that milk powders exposed to various relative humidities generally crystallize in the alpha-lactose hydrate form, but that spray- and freeze-dried lactose produced a mixture of the alpha hydrate and the beta anhydride.

Roetman & van Schaik (1975) showed that mutarotation takes place in solution, during processing and during storage in milk powders. The rate of mutarotation in the amorphous form depends on the sample temperature and the moisture content but mutarotation will also occur when the amorphous lactose undergoes crystallization. Roetman (1979) studied the relation between lactose crystallization and the structure of spray-dried milk powders. The particle porosity of "postcrystallized" products obtained by humidification during the agglomeration process is influenced by the partial crystallization of the lactose in the alpha hydrate form. This phenomenon may explain the significant improvement in the reconstitution of agglomerated skim-milk powders obtained by de Vilder (1986).

Some interesting results have also been obtained for whole milk powders. Wursch et al. (1984) evaluated the incidence of water content and storage temperature on the state of the lactose in milk powder. Lactose losses due to Maillard reaction accompanied by crystallization in the beta form was found in spray-dried whole milk powders with a water content of 3.1% stored at 60°C. The rate of these reactions increased with increasing moisture content and the transition to the alpha-lactose hydrate occurred at moisture contents higher than 6.1%. Saito (1985) studied the structure of instant skim-milk and whole milk powders particularly with respect to lactose crystallization during storage. If the crystallization of alpha-lactose hydrate occurred according to moisture uptake, only beta lactose was in the crystal state in whole milk powder stored at 37°C for 5 months at $a_w < 0.2$. He mentioned that the factors which initiated the specific crystallization may influence the physical properties of milk powders.

Loss of Available Lysine Due to the Maillard Reaction

Reaction mechanisms of non-enzymatic browning have been described in terms of browning or decrease of free amino groups (Eichner, 1975). Most of the deteriorative reactions affecting foods depend on water content. Two types of problems can be compared (Karel & Flink, 1983):

(a) deteriorative reactions for which a definite critical moisture content can be established, and
(b) reactions that proceed at all moisture contents but the rate of which depends strongly on the moisture content such as non-enzymatic browning.

Different authors have pointed out the need for kinetic models and kinetic data relating the rate of nutrient destruction to composition and environmental factors such as moisture content, water activity and temperature.

Labuza & Saltmarch (1981) gave a more extended review on the non-enzymatic browning reaction as affected by water in foods. They emphasized the need for a kinetic approach to better understand the effects of water content or a_w for both browning and protein quality loss in the Maillard reaction. The experiments described by different authors are discussed in relation to temperature, and activation energies are calculated from the available data given in the literature. For the browning reaction, the activation energies at various a_w levels in different food products generally fall in the range 20-40 kcal/mole but variations may result from differences in composition and physical structure of the complex food systems. For protein quality loss, water effects have also been examined, but in contrast to the browning reaction, which is generally considered a zero-order reaction, lysine loss has generally been considered to be a first-order reaction at least for up to 75% loss. The activation energy for lysine loss ranges from 10 to 38 kcal/mole but very different results could be obtained depending on how the available lysine is measured. Not enough results are described in the current literature to better determine the maximum of lysine loss as a function of a_w. Finot et al. (1981) described the furosine method as an indicator of the presence of lactose-lysine in milk. By this method it appeared that the loss of available lysine in milk products depended on the type of heat treatment. There was little or no loss of available lysine in freeze-dried or spray-dried milk powder, but up to 50% loss in roller-dried milk.

Interesting observations on the Maillard reaction have been obtained on casein-glucose models, in particular concerning the role of water. Studying the reaction between casein and glucose, Lea & Hannan (1949) observed that the rate of loss of amino-N was powerfully influenced by the a_w in the system, showing a maximum at 0.65 to 0.70 a_w for temperatures of 37°, 70° and 90°C. The marked falling off of reaction rate as the a_w was increased beyond 0.70 may be a dilution effect. Jokinen & Reineccius (1976) studied losses in available lysine during thermal processing of soy protein/glucose model systems. They obtained a regression equation which showed that the three variables most affecting the Maillard reaction and resulting in loss of available lysine were, in decreasing order of significance, glucose concentration, temperature, and a_w squared. Time had relatively little effect. In particular, at any given glucose level, the available lysine loss was greatest at an a_w between 0.65 and 0.70. In a model containing other constituents such as glycerol, Warmbier et al. (1976a) studied the kinetics of pigment production, glucose utilisation and loss of available lysine as a function of temperature, moisture content and water activity. Initial loss rate of both glucose and available lysine followed first-order kinetics. Large losses of available lysine occurred before brown discoloration was appreciable. A maximum rate at 0.4 to 0.5 a_w was found for the Maillard browning reaction which was considerably lower than the 0.65 to 0.70 a_w range found by Lea & Hannan (1949) and reported by Loncin et al. (1968). This downward shift in the a_w maximum for browning was interpreted as being due to glycerol addition which increased reactant mobility and solubility at lower a_w. Warmbier et al. (1976b) also determined the influence of the sugar/amine molar ratio on the kinetics of the Maillard reaction in the 0.52 to 0.60 a_w range for both pigment accumulation and available lysine content. The data showed that an increase in the lysine loss rate constant occurred as the glucose/lysine ratio was increased from 0.5 to 5.

Lysine loss and browning reactions in milk products from liquid milk to dried products including skim-milk and whey powders have been dealt with in some papers. Loncin et al. (1968) showed the loss of available lysine of milk powder kept at 40°C for 10 days as a function of a_w. With only four experimental points a maximum of reactivity was suggested at around 0.65 a_w. Ben-Gera & Zimmermann (1972) observed up to 88% of loss in available lysine for non-fat dry milk stored for 24 months at 40°C below 0.60 a_w with no indication of lactose crystallization.

Von Huss (1970) studied both crystallization of lactose and availability of lysine after storage of dried skim-milk powder at different a_w and investigated a connection between the alteration in the content of available lysine and the physical state of the lactose. He mentioned that moisture contents have to be differently interpreted according to whether the lactose is present in the amorphous or crystalline state, but no indications were given about the crystallized form. Whereas moisture contents of 4.9% to 6.8% only led to negligible storage damage of lysine when the lactose was present in the amorphous form, moisture contents of 5.1% to 5.8% already led to a marked deterioration of the available lysine after crystallization. In another paper, Von Huss (1974) concluded that the "the structural changes in dry skim-milk powders due to crystallization of the lactose considerably speed up the transformations of the Maillard type of reaction, but thereafter they fall off to a substantially lower rate again", and that "all the processes activated hereby are strongly dependent on temperature". Saltmarch et al. (1981) studied the kinetics of protein quality loss via the Maillard reaction in spray-dried sweet whey powders as a function of a_w, temperature and physico-chemical state of lactose. A maximum of the browning reaction appeared at 0.44 a_w, related to the fact that at this a_w, amorphous lactose began to shift to the alpha hydrate crystalline form with a resultant release of water, then acting to mobilize reactants for the browning reaction.

Kessler & Fink (1986) investigated the effects of heating and storage conditions on the chemical changes in liquid milk. The reaction kinetics showed that the loss of available lysine could be described by a second-order reaction which is valid over a wide range of temperature (from 4^{0} to 160^{0}) and time (from 1 second up to 1 year). An activation energy of 109 kJ/mole was found by regression analysis using the Arrhenius equation.

PRACTICAL EXPERIMENTS AND OBSERVATIONS

Our efforts in the Nestle Research Centre go in three directions:

(a) to better understand the mechanisms involved in the chemical and physical reactions;
(b) to improve methods for measuring fundamental parameters such as reactive lysine, water content, water activity and state of crystallization; and
(c) to develop and test mathematical models using our own data based on specific experiments.

Some topics in these directions are presented here.

Conditions to Produce Lactose Crystallization in Milk Powders

Crystallization of amorphous lactose in milk powders may occur under different conditions which can be clarified by the two following irreversible phase transition experiments:

(a) Uptake of moisture at constant temperature : By acting on the surrounding atmosphere, a stepwise increase in a_w can be imposed at constant temperature so that there is sufficient moisture uptake by the powder components to provoke a phase transition (Iglesias & Chirife, 1978; Karel, 1985). This may be the case in the classical procedure to obtain moisture adsorption isotherms if the final a_w assigned for instance by a saturated salt solution is higher than a critical value for phase transition at the temperature of the product. In this situation there is a water transfer from the saturated salt solution to the powder which could be reversed if crystallization occurs. Fig. 1 illustrates this situation in the sorption isotherm diagram (a), and from a kinetic point of view by weighing the sample during sorption (b).

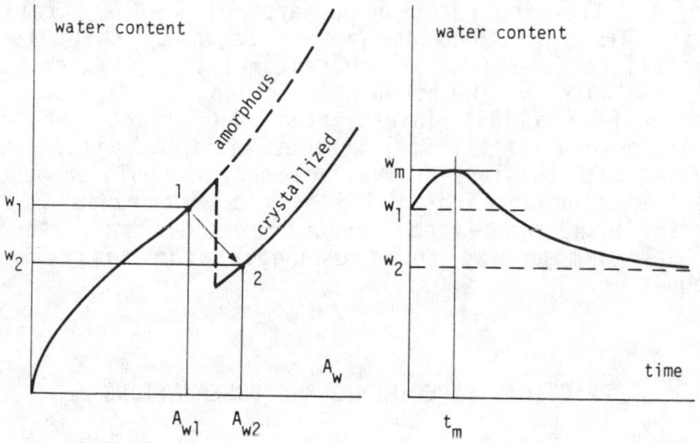

Figure 1. Lactose crystallization induced by water sorption at constant temperature.
a) sorption isotherm
b) kinetics of the change in total water content

(b) <u>Increasing the temperature at constant water content</u> : By heating the sample, an increase in temperature can be imposed at constant water content so that there is a sufficient increase in a_w to provoke the phase transition. This may be the case of a sample kept in a closed container for storage tests at different temperatures. Fig. 2 illustrates this condition in the sorption isotherm diagram (a) and from the kinetic point of view (b) by continuous water activity measurement. When crystallization occurs in this situation, water bound to the amorphous phase is either locally released and readsorbed by other constituents such as casein or chemically bound in the lattice when alpha-lactose hydrate is formed. As more water is available around the protein phase, the total effect of this rearrangement is a global increase of the water activity which may induce an acceleration of the Maillard reaction. Thermodynamic equilibrium can be long to achieve, which complicates the interpretation of kinetic models incorporating both crystallization and chemical reactions.

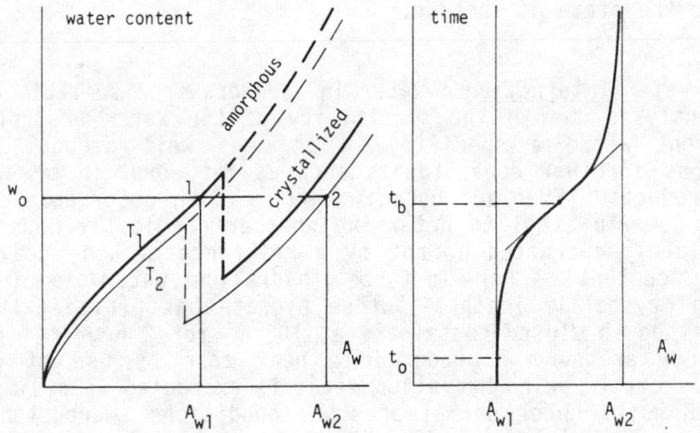

Figure 2. Lactose crystallization induced by increasing the temperature at constant water content.
(a) sorption isotherm
(b) kinetics of the change in water activity

In practice, when a product is kept in a non-hermetically sealed package at varying temperatures both of these phenomena may superpose to induce crystallization. In milk powders, the situation is complicated by the two possible forms of crystallization. We need an analytical method able to quantify the proportion of crystallized lactose and to distinguish between the two anomeric forms.

Near Infrared Reflectance Analysis of Lactose Crystallization

Near infrared reflectance (NIR) analysis is based on the principle of reflectance in the near infrared wavelength range between 1000 and 2500 nanometers. Spectroscopically it is the region lying between the electronic and the fundamental vibration bands of molecules. Absorption in this region arises mainly from C-H, O-H or N-H bonds, which characterize most food product constituents. The incident beam is selectively absorbed by the sample or diffused by back-scattering through an integrating sphere. The chemical composition could be correlated by mathematical algorithms to the absorbances measured at various wavelengths. This quantitative analytical method is currently used to rapidly analyze moisture content, fat content, protein content or other constituents in solid or liquid food products. There are hundreds of publications on chemical analysis, but only little information on physical analysis such as transitions from amorphous to crystallized state in sugars. In this section, we will show how we use our InfraAlyzer 500 (from Technicon) for analysing water content and for evaluating the state of lactose.

(a) Water interaction : Water in powders is a suitable application for NIR analysis due to the sensitivity of the water absorption around 1.94 microns where a specific O-H peak is well decoupled from the interactions of other constituents such as fat, protein or carbohydrate in food products. Free and bound water are both observable so that NIR is also a powerful tool to determine whether chemically bound water has been completely extracted or not by a reference method used to measure the water content. For instance, hydration water in milk powders containing crystallized alpha-lactose hydrate has proved extractable by an oven method in closed containers at 102°C for 2 hours if the partial water vapour pressure is reduced to near zero by use of phosphorous pentoxide. Little or no hydration water is extracted in a simple oven at 102°C in 2 hours under normal pressure conditions (Warburton & Pixton, 1978). This fact, illustrated by NIR spectra collected immediately after drying by the two methods (Fig. 3), may explain the surprising differences in the water content reported by Henderson & Pixton (1980).

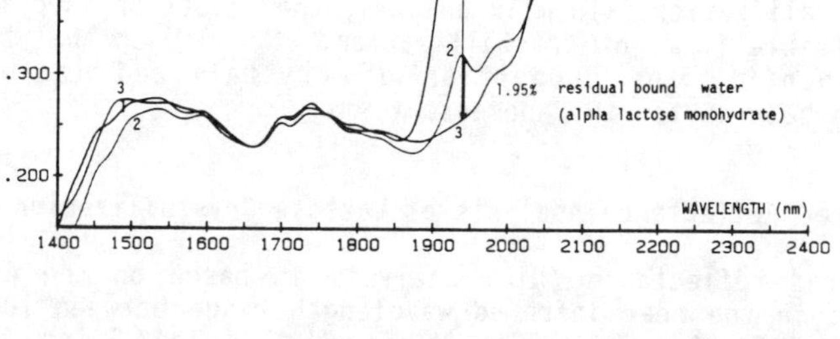

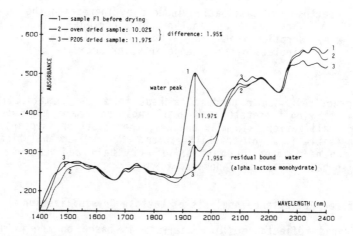

Figure 3. Comparison of the NIR spectra of a sample dried by the oven method and the same sample dried by the "P_2O_5" method.

(b) Crystallization interaction : Several authors describe physico-chemical methods for estimating the degree of lactose crystallization in dried milk products based on solubility and thermal properties (Sharp, 1938; Choi et al., 1951; Buma, 1970; Roetman, 1982). During our investigations, near infrared reflectance has proved of interest for directly observing the physical state of sugars in powders with the advantage of a rapid non-destructive method which does not require any special preparation of the sample. Thirty years ago, IR was already being discussed as a technique to test crystallization in sugars. Norris & Greenstreet (1958) discussed a communication by Goulden (1956) stating that an interaction between protein content and sugar was demonstrable by infrared absorption spectra in the 7 to 15 microns range. They proved then that these interactions were due to lactose crystallization and obtained spectra of amorphous and crystalline lactose in the alpha and the beta forms. Goulden (1958) suggested that these results were of particular interest in the interpretation of the infrared spectra of dried milk samples to follow crystallization as it occurred in the "instant process" for instance. Bushill et al. (1965) used infrared spectra and X-ray diffraction patterns to identify the crystalline forms of the lactose in milk powders and proved by this means that amorphous lactose may crystallize as a mixture of the alpha hydrate and the beta anhydride when exposed to air of fairly wide range of humidities. Susi & Ard (1973) described a procedure by infrared spectroscopy for evaluating the presence of crystalline alpha-lactose based on the observation that alpha-lactose hydrate exhibits strong absorption bands around 15.9 microns, while amorphous lactose either in the alpha or the beta form results in an almost flat background. The methodology to calibrate an instrument requires a sample set which covers as wide a range of chemical composition as possible. Having collected spectra of whole milk powder with an InfraAlyzer 500 in the range of 1.5 to 5.5% moisture, we were surprised to observe an unexpected peak around 2108 nm for several samples containing more than 5% water. It was subsequently proved that this interaction was due to lactose crystallization in the beta form which had been induced by the storage conditions in closed boxes between 25° and 30°C before calibrating the instrument. Fig. 4 shows spectra of amorphous and crystallized whole milk powder samples. Such crystallization in the beta form has also been obtained for samples in the range of 4.5 to 5.0% moisture by increasing the temperature and measuring continuously the a_w in the cell of the Rotronic DT hygroscope. For example, a sample at 4.6% moisture content and 28% fat was kept for two years without crystallization. By increasing the temperature to 50°C, the a_w increased from 0.42 to 0.61 in 48 hours. The bending of the recorded a_w curve was observed after 12 hours. The spectra before and after the heat treatment are shown in Fig. 5.

We have also measured the spectra of pure lactose in the different physical and chemical structures (Fig. 6). Amorphous lactose gives smooth spectra which are influenced by the texture but not by the beta/alpha ratio. Crystallized lactose shows peaks with specific interactions which are different for the alpha hydrate and the beta anhydride forms. In milk powder, other constituents partially mask these interactions but each form of crystallized lactose can easily be distinguished by the following observations:

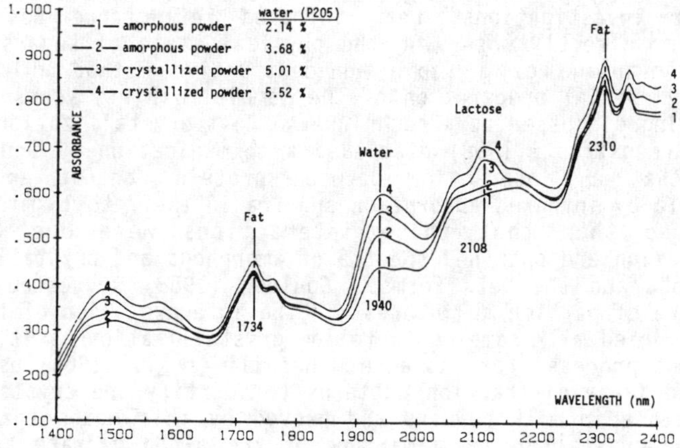

Figure 4. NIR spectra of amorphous and beta anhydride crystallized whole milk powders.

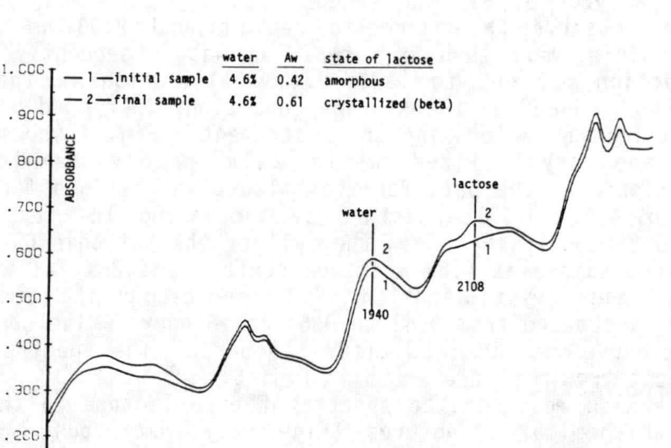

Figure 5. NIR spectra of an amorphous whole-milk powder and of the crystallized product after a thermal treatment at 50°C.

1) Peak around 2100 nm

- alpha lactose hydrate presents a peak at 2096 nm
- beta lactose anhydride presents a peak at 2108 nm

2) Peak around 1940 nm

- alpha lactose hydrate presents a narrow peak at 1936 nm due to hydration water
- beta lactose anhydride presents a small and wide peak due to the lack of hydration water and the small amount of adsorbed water

3) Doublet around 1500 nm

- alpha lactose hydrate presents a peak at 1540 nm
- beta lactose anhydride presents a more intense peak at 1484 nm as the amorphous lactose.

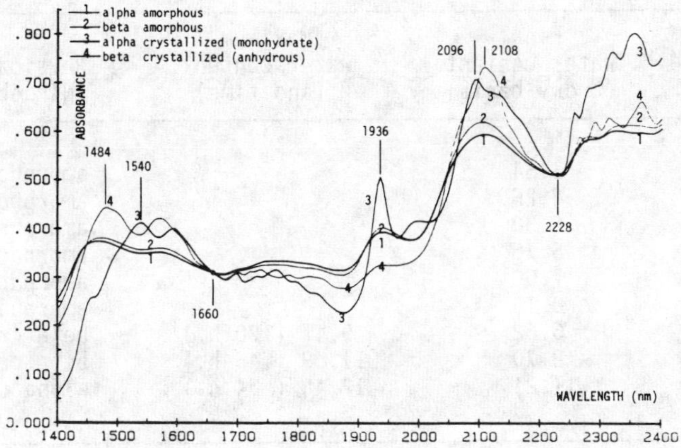

Figure 6. NIR spectra of amorphous and crystallized pure lactose.

We currently use these observations to evaluate the physical state of lactose in milk powders. The calibration procedure by linear regression analysis has not yet been developed because there is no other accurate method able to quantify each form of crystallized lactose in order to correlate the chemical values with the spectral data. Observations by polarization microscopy also give instantaneous information on the crystallization state but it is neither possible to distinguish the crystalline form nor to estimate the degree of the crystallization. Crystallization initiated at the surface of the particles gives the same image as the complete crystallization of the powder. Thus we used polarization observations to test the onset of crystallization, and NIR to test the crystallization form and to evaluate the degree of crystallization.

Sorption Isotherm of Skim-Milk Powder

We commonly determine sorption isotherms of various products by using a standard equipment similar to that recommended by the conclusion of the "Cost Project" (Wolf, 1985). Table 1 summarises the results of maximum and final moisture content after adsorption of skim-milk powders at an initial moisture content of 2.3% for various water activities. The state of lactose was determined by measuring the NIR spectra at the end of the sorption procedure.

TABLE 1
Water content and physical state of the lactose in skim-milk powder at various a_w levels.

a_w	Final water content (% dry basis)	Maximum water content (and time)	Physical state (NIR observ.)
0.113	3.54		amorphous
0.176	4.25		amorphous
0.230	5.30		amorphous
0.251	5.76		amorphous
0.304	7.00		amorphous
0.346	8.15		amorphous
0.391	6.62	9.47 (120 h.)	beta crystal
0.493	8.70	11.89 (48 h.)	beta crystal
0.572	11.21	12.40 (24 h.)	alpha crystal

Fig. 7 shows the kinetics of moisture uptake measured over more than 500 hours on about 1 gram of skim-milk powder. Here,

- from 0.11 to 0.35 a_w, the kinetic curves correspond to typical adsorption of amorphous products;

- at 0.39 a_w one observes the release of water after a long delay of 200 hours and equilibrium has not been reached after 500 hours (3 weeks);

- the crystallization rate is more rapid at higher a_w (0.49 and 0.57);

- the beta crystallization form is produced at 0.39 and 0.49 a_w, but it changes to the alpha form at 0.57 a_w (Fig. 8).

It can be concluded that the rate of crystallization depends on the available water which is minimum at a water activity near the breakpoint of the adsorption isotherm. Our results are similar to those obtained by Henderson & Pixton (1980). The anhydride crystallized form is produced so long as there is not sufficient water to produce the hydrate form.

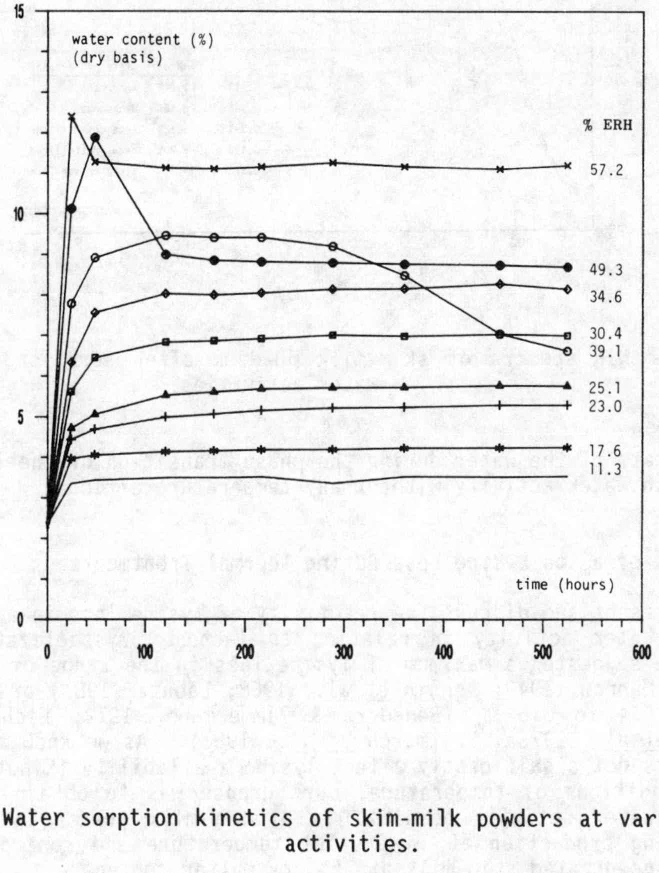

Figure 7. Water sorption kinetics of skim-milk powders at various water activities.

Fig. 8 shows the details of NIR spectra of samples equilibrated at various a_w and the identification of the crystallization state of the lactose after 520 hours. These spectra correspond to the first and the last four samples of Table 1. Fig. 9 illustrates the sorption isotherm measured at 25°C. One can conclude from this experiment that the state of the crystallized lactose changes with the final a_w. Around 0.40 a_w it is in the beta anhydride form and around 0.60 a_w it is in the alpha hydrate form for the same temperature of sorption. This means that the

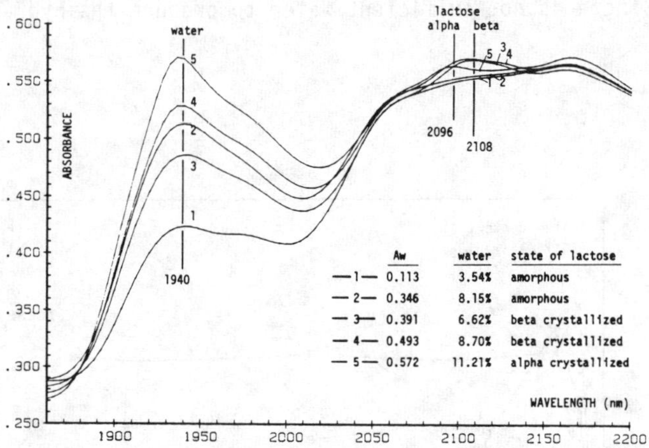

Figure 8. NIR spectra of skim-milk powders after sorption at various water activities.

binding state of the water during the phase transition of the lactose may change with water activity without any temperature effect.

The Effect of a_w on Lysine Loss During Thermal Treatments

Here we present and discuss the reactivity of lysine loss in a particular range of water activity in relation to lactose crystallization. The literature suggested a maximum of lysine loss in the range of 0.6 to 0.7 a_w (Lea & Hannan, 1949; Loncin et al., 1968; Labuza, 1981) or in a lower range of 0.4 to 0.6 a_w (Ben-Gera & Zimmermann, 1972; Eichner, 1975; Warmbier et al., 1976a; Saltmarch et al., 1981). As we know that spray-drying does not significantly affect lysine availability (Finot, 1983) in normal conditions of temperature, our purpose was to obtain the higher range of water activity possible in a skim-milk powder by a special spray-drying production at low outlet temperature and controlled inlet flow of concentrated skim-milk at 35% dry matter content.

(a) Experimental approach : Sticking, caking and crystallization were the major problems to take into account (Downton et al., 1982). The process conditions were determined during preliminary trials and a procedure was organized to rapidly freeze the samples in small bags with liquid nitrogen to block the physical state of the lactose. The powders were produced in two final trials from low to high moisture content by increasing the inlet flow in steps until either too much sticking

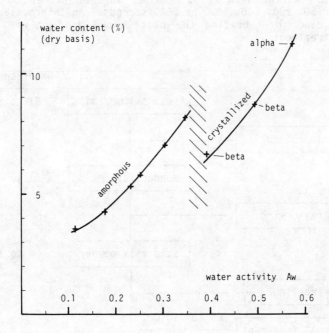

Figure 9. Sorption isotherm of skim-milk powder at 25°C.

occurred or crystallization was observed by polarization microscopy in the outlet product. The moisture content was controlled by an InfraAlyzer 400 (routine NIR instrument) calibrated in this range during the preliminary trials.

In the following discussion we will especially consider lysine loss due to thermal treatments and measure the change in water activity throughout thermal treatment in relation to lactose crystallization. Fig. 10 shows the flowchart of the process which can be divided into four steps:

- Evaporation : dry matter, fat, protein and initial lysine contents were measured before and after the evaporation step.
- Spray-drying : moisture content was monitored by the InfraAlyzer 400. The state of the lactose was controlled by polarization microscopy.

- Freezing : by liquid nitrogen and keeping the samples in a freezer at $-50^{o}C$ before and after the thermal treatments to avoid crystallization.

- Thermal treatments : by immersion of thin bags (200 x 100 x 5 mm) of powder in a water bath at 65°, 80° or $98^{o}C$ or in an oil bath at $110^{o}C$ for 50, 200, 500, 1000 or 3600 seconds. An intermediary bath at $0^{o}C$ was used for heating the bags before and cooling after the thermal treatments.

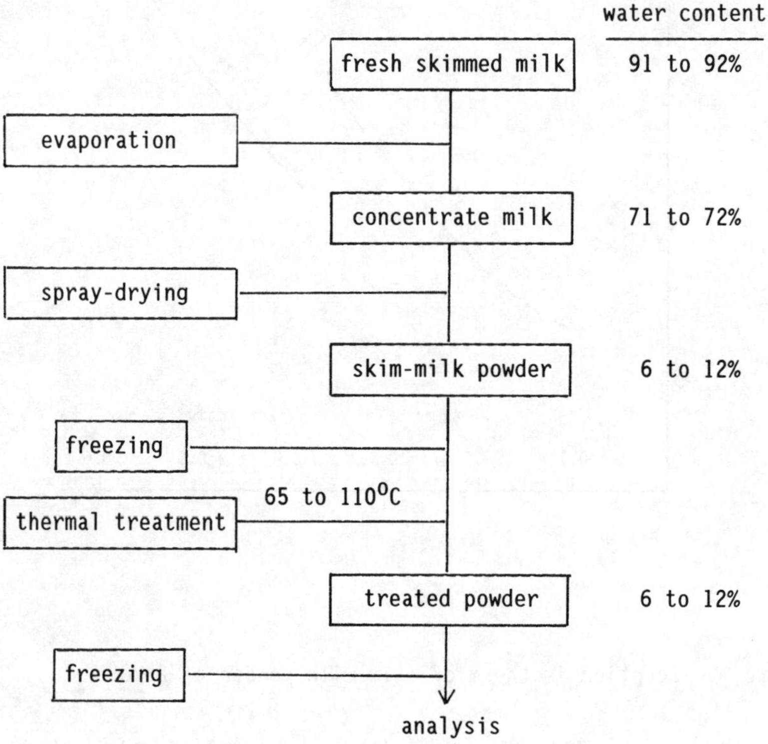

Figure 10. Flowchart of the process to obtain skim-milk powders at high a_w in the amorphous state.

The following analytical methods were applied:

1) water content

- classical oven method at $102^{o}C$ for 2 hours.
- special method at reduced total pressure (50 torr) and very low partial water vapour pressure by using P_2O_5, at $102^{o}C$ for 2 hours. This method is referred to as "P_2O_5" in the remainder of this report.

2) water activity

- determination at 25°C with a Rotronic hygroscope DT thermostatised and calibrated in the range of interest by using the LiCl non-saturated standard solution from Rotronic.

3) crystallization state

- observation of suspensions of powder in Canada balsam with the polarization binocular
- observation of the NIR spectra collected with the InfraAlyzer from Technicon.

4) lysine

- furosine method as described by Finot (1981): total lysine (TL) and furosine (F) were determined by chemical analysis.
- reactive lysine (RL) was calculated by the empirical relationship: RL = TL - 1.24 F
- blocked lysine (BL) in percent was calculated by the emprical relationship: BL = 100(3.1 F)/(TL + 1.86 F)

(b) Results and discussion : Six levels of water content were obtained covering the range from 0.29 to 0.58 a_w without observing crystallization of lactose. The last value was the limit for obtaining product at the outlet of the spray-drier. At higher water contents sticking and crystallization occurred. Table 2 gives the corresponding values of the water content determined by the two methods (wet basis), the total water content on a dry basis and the water activity. Fig. 11 illustrates the relationship between a_w and total water content as in the sorption isotherm. The curve can be equated to a desorption curve and covers the ordinary range where the breakpoint due to lactose crystallization is observed in the usual adsorption procedure. The differences between the water content determination of the two methods (0.58 +/- 0.07) are due to the fact that the "P_2O_5" method extracts more water than the oven method. This difference may increase for samples which contain alpha-lactose hydrate (see Table 4 and Fig. 3). Forty-five (45) samples including the six initial samples and 39 treated samples were analysed for water content, water activity, crystallization state, reactive lysine and blocked lysine. Many interesting observations can be made from the analytical results:

1) Water content:

a) There was a production of water in some samples during the longer or the higher thermal treatments. A significant production of 0.5 to 0.8% water was measured for all the samples treated at 98°C for 1000 s as summarized in Table 3. An increase from 6.01 to 8.66% water content was measured for the only sample which was submitted to a longer thermal treatment (3600 s at 98°C). As a little browning and a water production higher than expected by the "early" reaction (0.30% at 70% lysine loss) were observed, one can suppose that the "advanced" Maillard reaction had taken place during this treatment (Finot et al., 1981).

TABLE 2
Amorphous skim-milk powders obtained by spray-drying

Sample ident.	Water content (% wet basis)		Diff. (%)	Water content (% dry basis)	a_w (25^oC)
	P_2O_5	oven			
AO	6.01	5.48	0.53	6.39	0.287
BO	7.06	6.54	0.52	7.60	0.356
CO	8.69	8.03	0.66	9.52	0.440
DO	10.21	9.66	0.55	11.37	0.519
EO	11.34	10.67	0.67	12.79	0.552
FO	11.91	11.39	0.52	13.52	0.579

TABLE 3
Production of water measured by the "P_2O_5" method in samples treated at 98^oC for 1000 s

Sample ident.	Water content (% wet basis)		Diff. (%)	Physical state
	Initial	Final		
A8	6.01	6.89	0.88	amorphous
B8	7.06	7.83	0.77	amorphous
C8	8.69	9.36	0.67	amorphous
D8	10.21	10.81	0.60	beta crystal
E8	11.34	11.77	0.43	beta crystal
F8	11.91	12.40	0.49	beta crystal

b) The difference between the "P_2O_5" and the oven method strongly fluctuates depending on the state of the lactose. All the samples in the amorphous state showed differences in a narrow range of 0.3 to 0.7%. All partially or entirely crystallized samples showed differences between 0.4 and 2.0%. The samples crystallized in the beta form showed the same order of differences as the amorphous samples. Samples containing alpha-lactose hydrate showed higher differences up to 2.0%. Table 4 summarizes the differences for samples treated at 80^oC. This fact has been confirmed by the observation of NIR spectra of the dried samples (Fig. 3). It can be concluded that the water bound in the hydrate form is not extracted by the oven method.

TABLE 4
Differences between the water measured by the "P_2O_5" and the oven method in samples treated at 80°C for 1000 s

Sample ident.	Water content (% wet basis)		Diff. (%)	Physical state
	P_2O_5	oven		
B5	7.30	6.82	0.48	amorphous
C5	8.90	8.36	0.54	amorphous
D5	10.19	9.38	0.81	beta crystal
E5	11.31	10.01	1.30	alpha/beta
F5	11.90	10.31	1.59	alpha crystal

2) Water activity

All water activities measured after the thermal treatments were higher than the initial values. This increase can be due to different factors:

- lactose crystallization;
- production of water (Table 3);
- changes in texture due to heating and grinding the samples.

The distribution of the water was radically altered by lactose crystallization initiated during the thermal treatments. As the samples had been frozen at the end of the thermal treatment in order to block the transitory state of water, it was necessary to thaw them out before analyzing the different parameters. The water activity was measured half an hour after thawing at room temperature. For many samples the water activity was not found to be constant but an increase was observed, confirming that crystallization was not completed during the thermal treatment. The final constant water activity was reached only after some hours. This was also observed for the reference samples (without thermal treatment). For example, the water activity increased from 0.519 to 0.650 for the sample D0 and from 0.579 to 0.700 for the sample F0 which were both produced in the initial amorphous form.

Table 5 summarizes the increase in water activity in relation to the crystallization effect for the treatment at 80°C for 1000 s. One can conclude from these results that the mean increase of about 0.14 a_w is only due to the lactose crystallization, even if crystallization occurs only partially in this treatment. As the formation of hydrate liberates less water than the formation of anhydride crystals, the increase in a_w is lower for alpha (0.127) than for beta (0.151) crystallization.

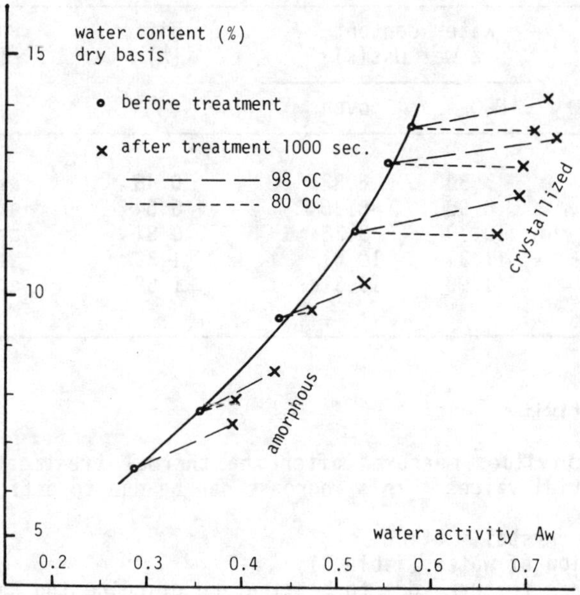

Figure 11. Relationship between the a_w and the water content of amorphous skim-milk powders.

3) Lactose crystallization

Analysing the crystallization effects by NIR spectra and according to the a_w measurements, two groups of samples can be separated:

- A, B and C samples at initial a_w from 0.29 to 0.44 do not indicate a significant degree of crystallization by NIR observation even if observations with the polarization binocular show surface crystallization.

- D, E and F samples at initial a_w from 0.52 to 0.58 exhibit various degrees and forms of crystallization:

  - at 0.52 a_w, beta crystallization occurs in most cases. There is only a little alpha hydrate formation at 65°C, 1000 s;

TABLE 5
Increase in a_w in relation with water production determined by the "P_2O_5" method and the crystallization state after the thermal treatment at 80°C for 1000 s

Sample ident.	a_w		Diff. in a_w	Water produced (%)	Physical state
	init.	final			
A	0.287	---	---	---	amorphous
B	0.356	0.393	0.037	0.24	amorphous
C	0.440	0.472	0.032	0.21	amorphous
D	0.519	0.670	0.151	-0.02	beta crystal
E	0.552	0.696	0.144	-0.03	beta/alpha
F	0.579	0.707	0.127	-0.01	alpha crystal

- at 0.55 a_w and 80°C, one can observe a phase transformation from amorphous to beta (200 s) and then to alpha (1000 s) forms. But this is different 98°C where there is first a transformation from amorphous to alpha (50 s) and then to beta (1000 s) forms;

- at 0.58 a_w, alpha crystallization occurs for the lower temperature treatments and beta for the higher temperature treatments.

This confirms that the alpha hydrate form is more stable at 80°C and the beta anhydride is more stable at 98°C (King, 1965).

4) Lysine losses

Blocked lysine up to 80% has been measured for the higher thermal treatments (1000 s at 110°C or 3600 s at 98°C). A difference between the initial reactive lysine of the group of samples C0 and E0 (7.64 g/16g N) and the group of samples A0, B0, D0 and F0 (8.37 g/16gN) can be explained by the fact that the groups have been produced at an interval of three months so that there is a difference in the protein content.

The second order kinetic model (Kessler & Fink, 1986) has been applied for the reactive lysine according to the equation:

$$RL(t) = RL(0)/(1+kt/RL(0)) \quad \text{where } t = \text{time (s)}$$

Thus $1/RL(t) - 1/RL(0) = k.t$

Kinetics of loss of reactive lysine using the relationship of this model are illustrated in Fig. 12. The "k" values were calculated by linear regression analysis of time vs the mean value of 1/RL(t) - 1/RL(0) for the different temperatures without taking into consideration the crystallization state. The same calculation was effected for both the amorphous and the crystallized samples. The ln(k) values were calculated in order to determine the activation energy by the Arrhenius law. The results are summarized in Table 6. They show that there is only a small effect of the crystallization state, and consequently of a_w.

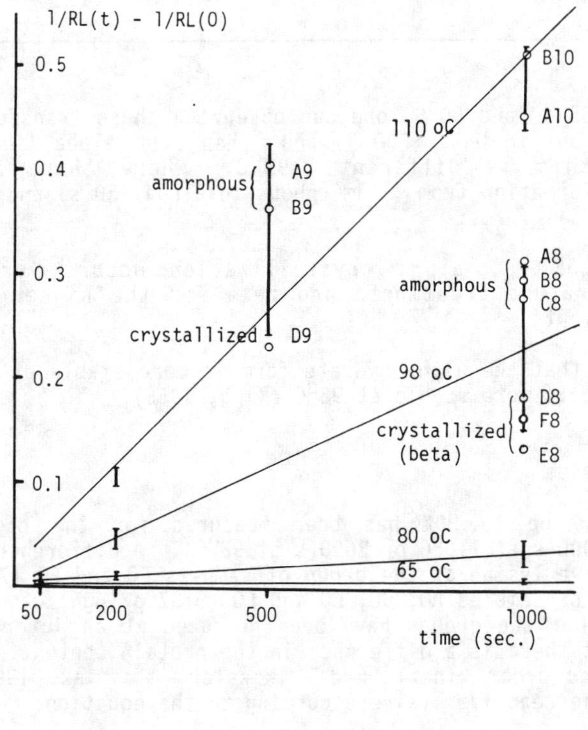

Figure 12. Reaction kinetics of the reactive lysine due to thermal treatments of skim-milk powders.

TABLE 6
Values of ln(k) calculated by the 2nd order kinetic model

Temp. (^{o}C)	1000/T ($1/^{o}K$)	All samples	Amorphous samples	Crystalline samples
65	2.96	-12.0	---	-12.0
80	2.83	-10.3	-10.0	-10.7
98	2.70	- 8.4	- 8.1	- 8.7
110	2.61	- 7.3	- 7.2	- 7.7

Comparing the ln(k) values for the amorphous and the crystalline samples one observes that crystallization results in a small deceleration of the reactive lysine reaction. But k_o and E_a values calculated by the Arrhenius equation are not significantly different. The activation energy E_a is in the range of 106 to 112 kJ/mole.

Fig. 13 shows the relationship between the blocked lysine for the different thermal treatments and water activity. As both lysine loss and lactose crystallization occur simultaneously in our range of experimental conditions one can expect that the two phenomena will interact. The high water activity leads to lactose crystallization which modifies the distribution of the total water. One part of the water can be bound in the alpha-lactose hydrate form but another part of the water is released and readsorbed by proteins. In a closed environment, lactose crystallization always induces an increase of the water activity of the skim-milk powder and this could accelerate the Maillard reaction. On the other hand, one can suppose that the lactose crystallization immobilizes a large part of the lactose so that less dissolved lactose is available for the Maillard reaction which will be slowed down.

One can conclude from our experiment that a_w has little or no effect on the loss of available lysine in the range of 0.3 to 0.6 a_w, but the maximum of reactivity has been attained as shown by a comparison with the results obtained in other experiments with dried milk powders at low a_w and with liquid milk at various concentrations at 80°C for instance. Liquid skimmed milks with 10 to 50% dry matter have ln(k) values from -15 to -13, skim-milk powders in the range of 2 to 4% moisture have ln(k) values of about -14 while skim-milk powders in the range of 6 to 12% moisture have a ln(k) value of about -10. The corresponding times to achieve 10% and 50% lysine loss, respectively, are given in Table 7.

CONCLUSION

Skim-milk is a complex but interesting product to study lactose crystallization and lysine loss. Lactose has been found to crystallize in the beta anhydride or in the alpha hydrate form depending on the water activity and temperature.

Near infrared reflectance can be used to evaluate the state of lactose, the degree of crystallization and to recognize the form of crystallization.

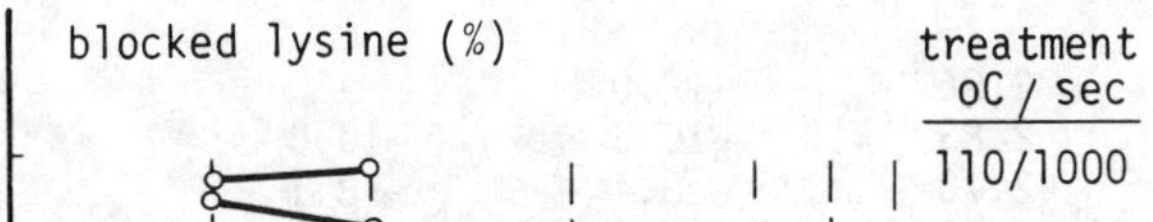

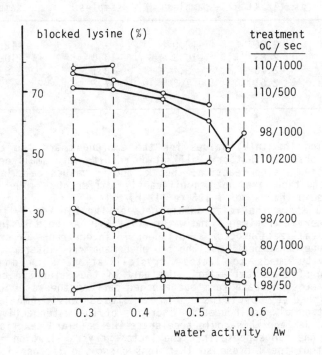

Figure 13. Relationship between the blocked lysine and a_w for different thermal treatments.

TABLE 7
Time to reach 10% and 50% lysine loss at 80°C in skim-milk at various dry matter contents

Parameter	Liquid 10-50% d.m.	Intermediate 6-12% water	Powder 2-4% water
ln(k)	-15 to -13	-10	-14
t(10% loss)	4 to 1/2 day	40 minutes	40 hours
t(50% loss)	38 to 5 days	6 hours	14 days

Water distribution and water binding are highly affected by lactose crystallization which induces an increase in the water activity. The continuous measurement of water activity in a closed box may be used to evaluate the kinetics of crystallization.

The method to determine the water content is critical. The "P_2O_5" method measures the total water content including the chemically bound water in the alpha-lactose hydrate, whereas the oven method measures most but not all of the total adsorbed water.

Skim-milk samples in the 0.3 to 0.6 a_w range may be produced and kept in the amorphous state by spray-drying and freezing.

The rate of lysine loss is maximum in the 0.3 to 0.6 water activity range. There is no significant effect of a_w or of crystallization on the loss of lysine in this range so long as the thermal treatments are short relative to the phase transformation kinetics.

ACKNOWLEDGEMENTS

This study is the result of a collaboration of many contributors from the Nestle company, from the Quality Assurance Laboratory and the Research Department of Nestec Ltd. I thank everybody who has suggested these trials, contributed to the production of the samples, discussed the model, made the chemical analyses, corrrected the manuscript and supervised the project.

REFERENCES

Ben-Gera, I. & Zimmermann, G. (1972). J. Food Sci. Technol. 9, 113.

Berlin, E., Anderson, B.A. & Pallansch, M.J. (1968). J. Dairy Sci. **51**, 1912.

Berlin, E., Anderson, B.A. & Pallansch, M.J. (1970). J. Dairy Sci. **53**, 146.

Buma, T.J. (1970). Neth. Milk Dairy J . **24**, 129.

Bushill, J.H., Wright, W.B., Fuller, C.H.F & Bell, A.V. (1965). J. Sci. Food Agric . **16**, 622.

Choi, R.P., Tatter, C.W. & O'Malley, C.M. (1951). J. Dairy Sci. **34**, 845.

De Vilder, J. (1986). Revue de l'Agriculture **39**, 865.

Downton, G.E., Flores-Luna, J.L. & King, J. (1982). Ind. Eng. Chem. Fundam. **21**, 447.

Eichner, K. (1975). In Water Relations of Foods (Duckworth, R.B., ed.), p. 417. Academic Press, London.

Finot, P.A., Deutsch, R. & Bujard, E. (1981). Prog. Food Nutri. Sci. **5**, 345.

Finot, P.A. (1983). Forschungsberichte **35** (3), 357.

Flink, J.M. (1983). In Physical Properties of Foods (Peleg, M. & Bagley, E.B., eds), p. 473. Avi, Westport, Conn.

Goulden, J.D.S. (1956). Nature **177**, 85.

Goulden, J.D.S. (1958). Nature **18**, 266.

Henderson, S. & Pixton, S.W. (1980). J. Stored Prod. Res. **16**, 47.

Hurrell, R.F. (1984). In Development in Food Proteins (Hudson, B.J.F., ed.), p. 213. Elsevier, London.

Iglesias, A.G. & Chirife, J. (1978). J. Food Technol. **13**, 137.

Jokinen, J.E. & Reineccius, G.A. (1976). J. Food Sci. **41**, 816.

Kaanane, A. & Labuza, T.P. (1985). J. Food Sci. **50**, 582.

Kanterewicz, R.J. & Chirife, J. (1986). J. Food Sci. **51**, 826.

Karel, M. & Flink, J.M. (1983). In Advances in Drying (Mujumdar A.S., ed.), vol. 2, p.103. Hemisphere Publ. Corp., Washington, D.C.

Karel, M. (1985). In Properties of Water in Foods in Relation to Quality and Stability (Simatos, D. & Multon, J.L., eds), p. 153. Martinus Nijhoff, Dordrecht.

Kessler, H.G. & Fink, R. (1986). J. Food Sci. **51**, 1105.

King, N. (1965). Dairy Sci. Abstr. **27**, 91.

Labuza, T.P. (1975). In Water Relations of Foods (Duckworth, R.B., ed.), p. 155. Academic Press, London.

Labuza, T.P. (1981). In Water Activity : Influences on Food Quality (Rockland, L.B. & Stewart, G.F., eds), p. 605. Academic Press, New York.

Labuza, T.P. & Saltmarch, M. (1981). J. Food Sci. **47**, 92.

Lea, C.H. & Hannan, R.S. (1949). Biochim. Biophys. Acta **3**, 313.

Linko, P., Pollari, T., Harju, M. & Heikonen, M. (1981). Lebensm. Wiss. u-Technol. **15**, 26.

Loncin, M., Bimbenet, J.J. & Lenges, J. (1968). J. Food Technol. **3**, 131.

Makower, B. & Dye, W.B. (1956). J. Agric. Food Chem. **4**, 72.

Norris, K.P. & Greenstreet, J.E.S. (1958). Nature **181**, 265.

Petriella, G., Resnik, S.L., Lozano, R.D. & Chirife, J. (1985). J. Food Sci. **50**, 622.

Roetman, K. (1979). Neth. Milk Dairy J. **33**, 1.

Roetman, K. (1982). Neth. Milk Dairy J. **36**, 1.

Roetman, K. & van Schaik, M. (1975). Neth. Milk Dairy J. **29**, 225.

Saito, Z. (1985). Food Microstructure **4**, 333.

Saltmarch, M., Vagnini-Ferrari, M. & Labuza, T.P. (1981). Prog. Food Nutr. Sc. **5**, 331.

Saltmarch, M. & Labuza, T.P. (1980). J. Food Sci. **45**, 1231.

Sharp, P.F. (1938). J. Dairy Sci. **21**, 445.

Supplee, G.C. (1926). J. Dairy Sci. **9**, 50.

Susi, H. & Ard, J.S. (1973). JAOAC **56**, 177.

Thomas, M.A., Turner, A.D., Abad, G.H. & Towner, J.M. (1977). Milchwissenschaft **32**, 408.

Von Huss, W. (1970). Landwirtsch. Forsch. **23**, 275.

Von Huss, W. (1974). Landwirtsch. Forsch. **27**, 199.

Warburton, S. & Pixton, S.W. (1978). J. Stored Prod. Res. **14**, 143.

Warmbier, H.C., Schnickels, R.A. & Labuza, T.P. (1976a). J. Food Sci. **41**, 528.

Warmbier, H.C., Schnickels, R.A. & Labuza, T.P. (1976b). J. Food Sci. **41**, 981.

Wolf, W. (1985). In Properties of Water in Foods in Relation to Quality and Stability (Simatos, D. & Multon, J.L., eds), p. 661. Martinus Nijhoff, Dordrecht.

Wursch, P., Rosset, J., Kollreutter, B & Klein, A. (1984). Milchwissenschaft **39**, 579.

STUDIES ON THE STABILITY OF DRIED SALTED FISH

K.A. BUCKLE, R.A. SOUNESS,
Department of Food Science and Technology
The University of New South Wales
P.O. Box 1, Kensington, NSW 2033
Australia

S. PUTRO
Research Institute for Fishery Technology
P.O. Box 30, Palmerah, Jakarta
Indonesia

AND

P. WUTTIJUMNONG
Department of Agro-Industry
Prince of Songkhla University
Haad Yai, Thailand

INTRODUCTION

Indonesia is a major fish-producing country with a projected maximum sustainable yield of about 5 million tonnes/annum. Of the current fish catch approaching 2 million tonnes, some 34% of inland fish and 48% of marine fish are cured, with 71% processed by salting and drying. In East Java more than 70% of fish is processed by salting and drying. These traditionally dried fish, which represent a low cost source of high quality protein, have an estimated 30% loss as a result of inadequate drying and associated deterioration during transport, warehousing and retail distribution. To achieve the Indonesian Government's aim of increasing per head consumption of fishery products from 12 to over 22 kg/year it is essential that handling, processing and distribution systems are improved and associated postharvest losses are reduced.

Such was the situation in 1983 when discussions were held between the Research Institute for Fishery Technology in Jakarta, part of the Agency for Agricultural Research and Development, The University of New South Wales, in Sydney, Australia, The University of Brawijaya, in Malang, Indonesia, and the International Development Research Centre, Canada, concerning a joint project aimed at reducing postharvest losses of dried salted fish with particular reference to East Java. In mid-1984, a joint project involving these Institutions was commenced with the financial support of the Australian Centre for International Agricultural Research.

The scope of this joint ACIAR-AARD Project was broad, and its aims were comprehensive in a technical sense. But foreign aid projects in developing countries, involving some aspects of technology transfer, require more than just the development of technological solutions to apparently technological problems. Fortunately, the originators of this project were able to see beyond the boundaries of food preservation, food spoilage, food dehydration and food packaging, and included a comprehensive socio-economic assessment of the effects of changed technology on those involved in this industry, i.e. from the fishermen who catch the fish, to the processors, the sellers and the buyers at the retail markets. A mass of data on food production and processing, as well as incomes and prices, have been obtained through Government and other sources, and will be used to assess whether changes in technology, involving food production and processing, will have long-lasting beneficial or undesirable consequences; and if so, who will be affected.

All the phases of fish capture, processing and marketing have been integrated into a single study in recognition of the fact that the current spoilage problem associated with dried salted fish can be attributed to failings in the handling, packaging, storage, transport and retailing of dried fish as much as to inadequacies in current processing and drying practices.

The economic assessment part of the Project has been conducted by La Trobe University School of Agriculture in conjunction with the Centre for Agricultural Economic Research in Bogor, Indonesia.

In an attempt to limit the scope of the Project to a manageable level, it was decided to focus the study on one major fishing village on the Bali Straits of East Java - Muncar was chosen because it represents a microcosm of traditional Indonesian fish capture and processing practices, and also because The University of Brawijaya and IDRC had already carried out some project work in this village on the production and spoilage of dried/salted fish, and contacts had been made with fish processors and Government fishery officers. Some 75 to 90% of the fish catch in the Muncar region comprises the oil sardine or lemuru (*Sardinella longiceps*), with smaller catches of other sardines, tuna, chub mackerel, scad, ponyfish, squid and others. The lemuru can have an oil content between 1 and 28%; the high oil content causes problems in drying and storage and sometimes limits the shelf-life. Until its virtual disappearance from the Bali Straits last year, our Project concentrated on lemuru processing because of its economic and regional significance in Muncar.

AIMS OF THE PROJECT

The original aims of the Project were to characterise and evaluate current traditional practices, their variations and defects, and to conduct experimental studies designed to establish solutions, from the time of fish capture to the time of retail sale of dried fishery products emanating from Muncar; and to accompany this study in all phases by an economic evaluation and cost/benefit analysis for procedures developed hopefully to overcome such defects. Like all such projects, some of these aims have been realised but others have not. The Project is still in progress and significant results are expected in the next 4-6 months.

The experimental procedure has been divided into 7 areas (Table 1). In each area it has been necessary to define the current traditional practices, and by experiment determine the effect of variables on the quality characteristics and economics of dried salted fish:

Area 1 - On-board handling (RIFT) : Traditional practices were defined, and a cost/benefit analysis of the use of chilled sea water (CSW) carried out.

Area 2 - Fish landing and classification (RIFT) : For traditional practices, RIFT determined tonnage, usage patterns, price of major species, chemical, sensory and microbiological grading factors; grade, end use and quality were determined for fish landed, especially using CSW.

Area 3 - Salting/brining (UNIBRAW, UNSW, RIFT) : Current traditional practices such as salt concentration, time, brine re-use and sanitation were examined, and the benefits of variation in salt concentration, time and pH on fish grade and drying characteristics were assessed.

Area 4 - Drying at Muncar (UNIBRAW, RIFT) : Current drying methods (time, salt content, a_w, moisture content, environmental conditions) and problems were determined.

Area 5 - Fundamental drying studies (UNSW) : Fish drying curves under a variety of conditions (air temperature, air velocity, RH, product orientation, fatty vs non-fatty fish, pre-drying treatments, fish shape and size) were established, and a low-cost fish drier suitable for the Muncar environment is under development.

Area 6 - Evaluation of dried products (UNIBRAW, RIFT, UNSW) : Some current traditional products were defined in sensory, chemical and microbiological terms, and the shelf-life determined, as affected by packaging and other variables.

Area 7 - Packaging, storage, etc. (RIFT, CAER, LTU) : The economics of production of dried salted fish by traditional technology will be compared to that for fish preserved by the developed technology, including changes to on-board handling, salting, drying, packaging, storage and distribution.

TRADITIONAL PROCESSING OF LEMURU IN MUNCAR, EAST JAVA

Processing of dried salted lemuru is carried out in the Muncar area by over 60 processors, including small (50-100 kg), medium and large (>10 t/day) operations. During the past 5 years the lemuru has comprised 76-88% of the total Muncar catch, of which a significant proportion is dried in 3 grades (good, medium, poor). In order of decreasing quality and auction price, lemuru are processed by canning; salting and boiling (pindang); salting and drying; or converted into fish meal (Table 2). The majority of lemuru are caught in about 300 purse seiners (170 in Muncar) and landed in Muncar or in neighbouring ports in Bali and trucked to Muncar for processing. After landing the fish are auctioned (but estimates are that 40% bypass the auction), then transported to the processor, where they are brined in saturated salt for 2-48 h (dry season, generally 24 h) or up to 4-5 days in the wet season. The fish are then spread onto racks to dry for 1-3 days or dried by the side of the road. From about 11 a.m - 2 p.m. the fish often are covered to prevent excessive scorching in the sun and loss of fat. After cooling overnight, the dried lemuru are boxed and shipped by truck to wholesalers (e.g. in Surabaya, Bandung) for ultimate sale in the markets. The largest market is in West Java.

PROBLEMS

There are a number of problems with the traditional practices of salting and drying lemuru:

* the lemuru season has proven to be rather unpredictable, with no clearly defined season, and landing of fish is confined to about 2 weeks/month since they are not caught about one week either side of a full moon

* brines for salting are reused up to 20 to 30 times resulting in high levels of microbial and insect contamination especially in the wet season, leading to highly salted fish and to excessive fly contamination of fish during processing; in recent years there has been a trend towards consumers preferring lower salt fish

* inadequate packaging during storage and transport results in high levels of microbial and insect contamination which ultimately reduces the quality of the final product offered for retail sale

* the lack of ice results in the landing of fish for auction in a poor condition which cannot be overcome by subsequent processing

* insecticides such as Baygon are used frequently, including being sprayed directly onto the fish

* hygiene during auctioning, brining, drying, packing, transport, storage and sale generally are poor, with numerous opportunities for microbial and insect contamination that leads to 'losses' of product, either from the point of view of nutrition, quality or economics.

Under **ideal conditions**, however, good quality fish can be produced, with salting overnight and drying sometimes within 1 day to an adequately low moisture content or water activity (a_w). Hence any artificial drying regime introduced should at least be able to produce dried salted fish of the quality possible under optimal traditional practices, but should also be capable of drying fish during the wet season when sun drying conditions are sub-optimal, using technology that is economically viable and technologically simple.

TABLE 1
ACIAR-AARD project on fish drying in East Java

Area	Study	Institution
1	On-board handling	RIFT
2	Landing, classification	RIFT
3	Salting, brining	RIFT, UNSW, UNIBRAW
4	Drying at Muncar	RIFT, UNSW
5	Fundamental drying studies	UNSW
6	Evaluation of dried products	RIFT, UNSW, UNIBRAW
7	Packaging, storage, transport distribution, marketing, retail	RIFT, CAER, LTU, UNSW

RIFT = Research Institute for Fishery Technology, Jakarta
UNSW = University of New South Wales, Sydney
CAER = Centre for Agricultural Economic Research, Bogor
LTU = La Trobe University, Melbourne

TABLE 2
Auction price vs quality grade of lemuru

Quality grade	Final use	1985 Auction price (Rp/kg)	Approx. 1987 prices (Rp/kg)
Good	Canning	100-300	300-500
Medium	Salted, boiled (pindang)	90-120	200-350
	Salted, dried	45-100	150-200
Poor	Salted, dried	45-100	150-200
	Fish meal	40-90	80-150

RESULTS TO DATE

Areas 1 & 2 - On-Board Handling, Landing, Classification

Before May 1983 no fishing boats in Muncar were using ice or chilled sea water. In 1983, RIFT constructed a CSW prototype using a round-bottomed purse seiner. The CSW fish tank was similar to a simple insulated cool box with the bottom following the contours of the boat (Putro, 1986). Since that time 84 purse seiners at Muncar have had facilities constructed for storage of fish in CSW. The result has been a marked increase in both fish quality and auction price, with higher proportions of fish suitable for canning. Higher quality fresh fish, however, does not necessarily mean better quality traditional dried salted fish - experiments with lemuru of variable initial quality resulted in dried salted fish of fairly similar quality when processed by traditional techniques, because of the inherent defects in the traditional system. Some work has indicated that good quality dried salted lemuru is best obtained from fish that are not particularly fresh.

What still needs to be ascertained is whether good quality (CSW) lemuru, if salted and dried under better conditions (including artificial compared to sun drying), can produce dried salted fish that is preferred, or at least equally liked, and if so, what are the comparable costs incurred and prices obtained.

The introduction of CSW technology is a good example of how a change in technology can be introduced, not by outside or even local scientists telling the local fishermen what they should do, but by letting the fishermen see for themselves the benefits that such technology can bring - in this case, higher prices and better quality fish at auction. The spread of such technology occurs by word of mouth, and by seeing that competitors have an advantage with this technology.

Area 3 - Salting and Brining

A number of studies have been conducted on the salting or brining of fresh lemuru in Indonesia; of frozen lemuru in Australia; and of fresh Australian sardines (Sardinops neopilchardus) in Australia.

Initial saturated brine pH (4.0-9.4) and brining temperature (25°-35°C) had no significant effect on either salt uptake or moisture loss of low-fat (2.8% wet basis) Australian sardines. The final brine pH was approximately the same (pH 5.8-5.9) after 68 h salting.

Several studies examined the effect of brine pH on the salting behaviour and storage stability of lemuru processed in Indonesia. Lime juice (10%) and citric acid-containing brines (pH 2) produced dried salted fish of acceptable or superior quality compared to the traditional product, with no microbial spoilage. Tamarind pulp (25% in the brine) produced dried fish of acceptable odour and taste but the appearance was unacceptable. Further studies by RIFT on the use of traditional acidifiers (e.g. Belimbing wuluh) indicated no significant impact on salting and drying behaviour; but the storage stability of these treated fish compared to traditional fish still remains to be confirmed. Studies are continuing with fish treated with acetic acid (e.g. by spraying) before storage.

The use of acid, if not added to the salt brines then added by dipping or spraying the fish before the completion of drying, is considered worthy of further study, since a number of observations of salted fish during drying at Muncar have indicated significantly reduced fly landings on fish treated with sprays of 10% acetic acid, particularly towards the end of drying. Since fly and other insect infestation of fish during salting, drying and distribution represents a considerable cause of subsequent wastage, simple surface acidification may help to reduce insect contamination, especially if such a treatment can be coupled with improved packaging during distribution to provide a physical barrier to further attack. Acidified fish samples, incidentally, were invariably rated as good or better than the control dried fish. Studies by one of our colleagues in Bali, Indonesia, has shown that acid treatments of other fish improved their stability and organoleptic acceptability.

The salt uptake was somewhat slower in lemuru (fat content 12.8%) than in Australian sardines (fat content 6.5%) when salted at 19-22°C in saturated brine. Salting models for whole fish brined at ambient temperatures (19°-22°C) were developed (based on models of Zugarramurdi & Lupin, 1977, 1980) that produced fairly good agreement with experimental data for whole Australian sardines and lemuru (Fig. 1). The more that is known about the salting of such fish, the better will be such modelling.

Although the microbiological quality of salt brines reused many times deteriorated markedly, there appeared to be little effect of this on the organoleptic quality of salted fish that were dried under ideal conditions. However, in poor weather when the moisture content and hence a_w remained high, subsequent storage life was reduced if fish were excessively contaminated. Excessive re-use of brines is undesirable especially since it tends to select for the microorganisms that are best able to grow on dry salted fish.

Area 4 - Drying at Muncar

Studies on the drying of lemuru at Muncar have been hampered severely by the virtual disappearance of the fish since early 1986. No satisfactory explanations have been forthcoming, although severe overfishing in recent years probably has been a major contributing factor. In 1976, of 67000 t of fish were processed in Muncar, including 57000 t of lemuru, but since then there has been a dramatic reduction in the quantity of fish processed. Only about 1000 t was landed in Muncar in 1986.

Area 5 - Fundamental Drying Studies

A number of studies have examined the factors important in the drying of sardines. These studies have been assisted by the construction at UNSW of a mechanical drier with control of temperature to 60+0.5^{o}C, RH (ambient - 100%+2-5%), and air velocity (0.2-7 m/s). Weight of fish during drying is determined <u>in situ</u> and results can be fed by data logger into a computer to enable drying curves to be prepared immediately following a drying run. The flexible ductwork enables drying rates as affected by fish orientation to be determined.

Studies on Australian sardines (fat content 1-1.5%) salted for 24 h in saturated brine at 25°C, then dried at 35°C, 45°C, and 55°C at RH corresponding to heated average Denpasar air (28°C, 83% RH, i.e. 50%, 30% and 18% RH) showed no constant rate drying period; no significant

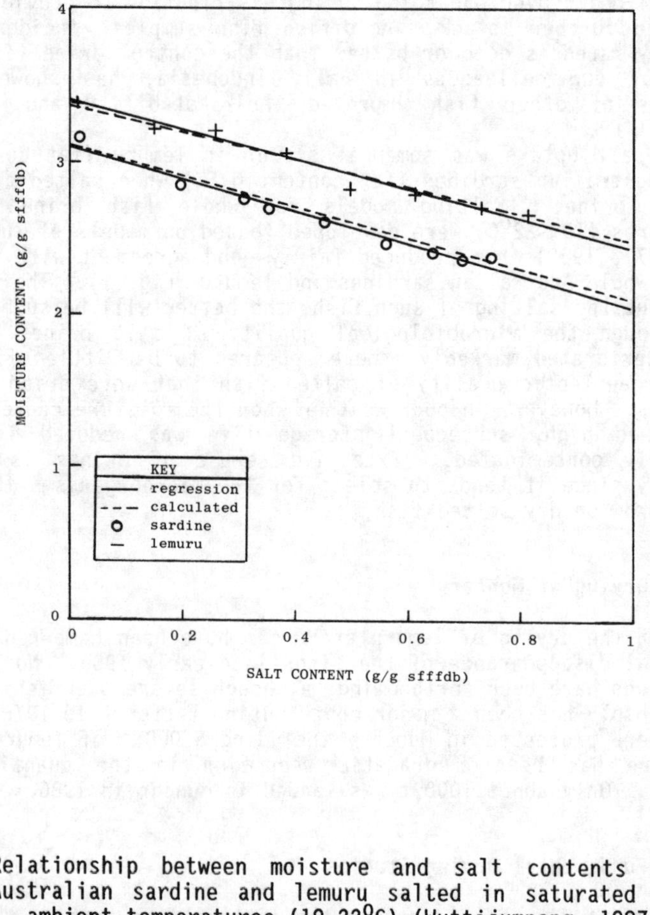

Figure 1. Relationship between moisture and salt contents of whole Australian sardine and lemuru salted in saturated brine at ambient temperatures (19-22°C) (Wuttijumnong, 1987).

effect of air velocity (1.0 vs 3.0 m/s) or fish orientation on drying rate; and a drying temperature of 55°C was detrimental to dried fish quality, i.e. the fish were brittle, case-hardened, and the skin was easily peeled. A drying temperature of 45°C gave the best quality dried fish under these conditions.

In the absence of high-fat fish, other studies were conducted on low-fat Australian fish. Australian yellowtail (Trachurus mccullochi), fat content 0.9%, were salted in 15%, 21% and saturated salt brines, then dried at 35°C and 50% RH, 45°C and 30% RH, and 55°C and 18% RH. Samples were assessed for moisture content, a_w, salt content and sensory properties. The lower the brine concentration, the longer the fish took to reach equilibrium for salt content or moisture content, but fish that had been brined in 15% salt reached a lower final a_w and moisture content than did fish brined in saturated salt, with fish brined in 21% salt falling between the other two. Similar studies subsequently with frozen lemuru and fresh Australian sardines processed in Australia gave similar results, i.e. brining in 15% salt produced dried fish of lower a_w than when fish were brined in saturated salt (i.e. the traditional method). Organoleptic assessment showed that fish salted in 15% brine and then dried at about 45°C were the most acceptable, possibly because of the lower salt content which panelists stated was preferred. This salting and drying regime needs further investigation.

We have also carried out studies on computer modelling of the drying behaviour of Australian sardines. Two models, a bimodal diffusion model and a two-term approximation of the diffusion equation model were fitted to the experimental drying data for _Sardinops neopilchardus_ dried at 35-55°C, 40-60% RH, 0.5-2.5 m/s, salt content 0.15-0.35 g/g salt-free, fat-free dry basis, using the SHAZAM package; the predicted moisture contents were in reasonable agreement with the experimentally determined values for this fish species (Fig. 2). This model, however, is applicable only to the Australian sardine, and further work is required to develop a model that is applicable to the lemuru or other commercially important fish species in S.E. Asia. Of particular importance in such a model would be the inclusion of the effects of fish size and fat content (Wuttijumnong, 1987).

Rice husk-fired furnace and fish drier : The traditional system for producing dried salted fish in Muncar works well providing conditions are optimal. When they are not, product quality deteriorates through mechanical damage or mould or insect attack resulting from inadequate packaging or inappropriate levels of moisture and/or salt. Any mechanical or non-sun drying system that is used to supplement or perhaps replace traditional sun drying ideally should conform to the following requirements: low capital cost (= short payback period), low running cost, manufactured from local materials, constructed by local tradesmen, minimal maintenance, repaired by local tradesmen, and minimal additional expenditure (e.g. labour) over the traditional process.

For the past year we have experimented with a rice husk-fired burner and natural draught drier that could be used to dry fish to supplement or replace traditional processes. It is based on a used 200 L steel drum with a centre tube containing 20-30% free-space as holes (Fig. 3). The drum outside the centre tube is filled with rice husks that are ignited initially by a wood or gas fire until pyrolysis begins. The gases produced during pyrolysis are released into the centre tube and ignite in a secondary combustion, producing flue temperatures of 800-1100°C. The hot flue gases are drawn through the walls and floor of a drying chamber and are released through a vertical flue - the released warm air draws ambient air through vents in the base of the drying chamber to produce the desired air temperature for drying. A 200 L furnace can produce an average 11 kW, with a peak of 20 kW over a 4-5 h period, and refilling while in operation should extend the run to 7-8 h; 27 kg of rice husks should dry about 150 kg of salted fish in 8 h.

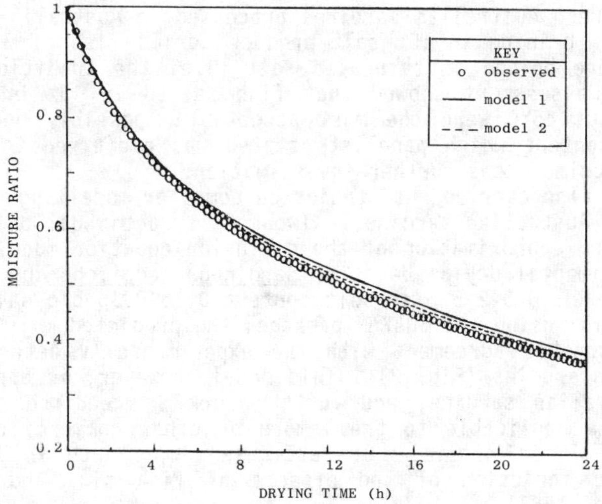

Figure 2. Comparison of moisture ratio predicted by the bimodal diffusion model (model 1) and the two-term model (model 2) with experimental drying data for salted Australian sardine (Wuttijumnong, 1987). Air temperature 45°C, relative humidity 50%, air velocity 1.5 m/s, salt content 0.15 g/g (salt-free, fat-free dry basis).

During May 1987 a furnace was built in 2 days at RIFT, Jakarta, from local materials, and a basic version should be able to be built in a village for about $A35. We have recently constructed a drying chamber in Australia, and a complete unit will be built in Jakarta in October 1987, with an approximate cost of $A60-80, again using local raw materials (e.g. mud and corrugated iron).

Although we have used rice husks, a number of agricultural wastes are possible sources of fuel. Agrowastes generated in Indonesia include rice straw (0.9 Mt coal equivalents), rice husks (3.6), wood waste (1.2), sawdust (0.9), coconut husk (6.8) and bagasse (2.3), giving a total of 15.7 Mt coal equivalent.

The design is simple, there are no moving parts, and a variety of locally-available or readily available raw materials could be used. A medium size dried salted fish processor in Muncar has volunteered to instal and run a drying unit in parallel with traditional drying of fish to enable any technological problems and cost/benefit analyses to be carried out. The results will be known early in 1988.

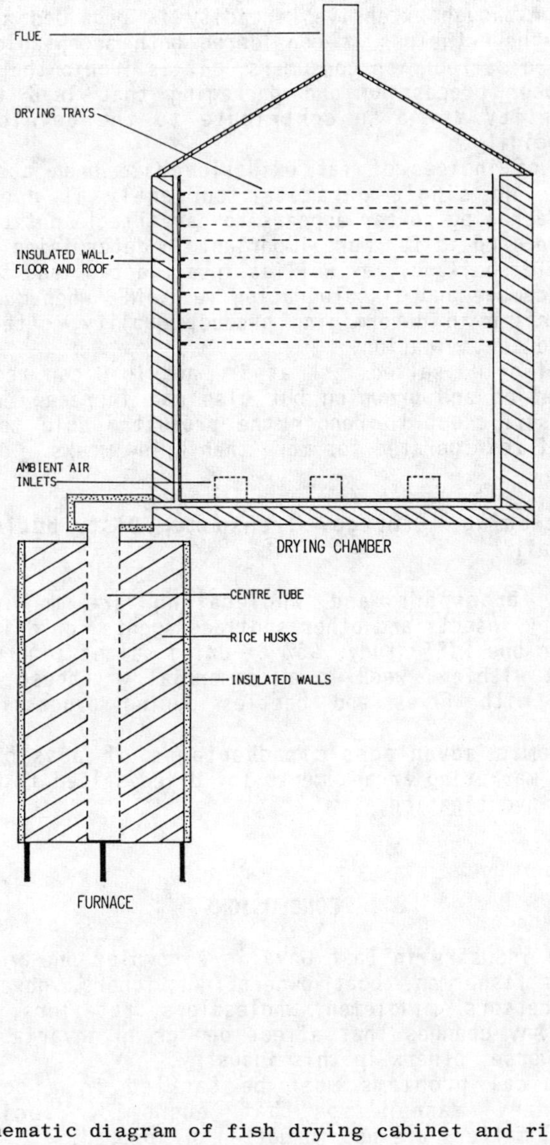

Figure 3. Schematic diagram of fish drying cabinet and rice husk furnace being developed at the University of New South Wales.

Area 6 - Evaluation of Dried Products

Product available in the market place is of tremendously variable quality, even within each of the 3 common grades, with considerable overlap between grades. A grading system has been developed, that includes assessment of appearance (physical damage, sheen, discolouration), texture, and infestation with mould, insects and pink bacteria.

The situation regarding rancidity of dried salted fish is rather interesting. Although extensive rancidity is regarded as unacceptable, some rancidity nevertheless is considered both acceptable and necessary by regular dried salted fish consumers. It is ironic that the conditions of hygiene, poor processing and packaging that lead to considerable losses of quality may also contribute to the development of some acceptable rancidity.

A number of indices of fat oxidation have been assessed in dried salted fish. No single parameter currently is useful under all conditions. Packaging so far appears to have little noticeable effect on lipid oxidation, and is far more important in determining moisture uptake (and hence mould spoilage), as well as insect attack and texture.

Physical damage and discolouration (e.g. NEB when muscle dries out) are very important in determining product quality, often more so than insect or mould contamination.

Storage of dried salted fish at 5°C and 10°C can reduce the extent of lipid oxidation and browning but also can increase moisture content and hence a_w sufficient to render the product liable to spoilage from mould growth if refrigerated for more than a few weeks.

Area 7 - Packaging, Storage, Transport, Distribution, Marketing, Retail

Distribution, transport and wholesaling are major sources of contamination by insects and other spoilage agents for dried salted fish. For example, in one RIFT study, 93% of dried salted lemuru were infested with an insect within 1 week in a commercial warehouse, and about 20% were infested with flies and beetles during processing but before marketing.

The economic advantages/disadvantages of possible changes to technology and marketing arrangements for dried salted fish are currently under intense investigation.

CONCLUSIONS

The dried fish industry in East Java is a complex one with many people involved, from fishermen, boat owners, auctioners, government fishery officials, processors, middlemen, wholesalers, retailers, consumers, and many others. Any changes that affect one group invariably affect, for better or for worse, others in this industry.

Technological problems must be tackled in a cooperative and multidisciplinary manner so that economic, sociological and anthropological matters are not forgotten or ignored.

CSW clearly is beneficial for fish quality and for returns for fishermen and boat owners, but not necessarily for the quality of dried salted fish. The way it was introduced perhaps indicates the way other changes to traditional processing will also be introduced. The long term effects of CSW on the dried salted fish industry or pindang industry still remain to be seen.

Much more can be done in conjunction with the processors to optimise the production of good quality traditional sun-dried salted fish (e.g. acid spraying and adequate packaging). If hygienic conditions during fish salting, drying and distribution are good, product contamination and hence product loss are reduced, as demonstrated by the new fish drying areas at Muara Angke near Jakarta. In the absence of such good conditions, processors can rely only on control of salt and moisture, and prevention of recontamination.

It is clear that any significant changes to current practices will have to include:

* use of CSW for those processing fish that do benefit from better quality at landing

* optimisation of both salting and drying practices to achieve maximum a_w reduction in the minimum time

* the use of a drying process that can occur irrespective of the ambient conditions, i.e. when conditions are poor

* the use of pre- or post-drying treatments, perhaps involving acid, that discourage insect proliferation and microbial spoilage

* improved packaging, perhaps only a plastic or paper lining in a traditional box, that will provide an additional barrier to moisture uptake and insect contamination.

The final stages of the Project will determine whether such a scenario is both technically and economically feasible.

ACKNOWLEDGEMENT

The authors acknowledge the financial support of the Australian Centre for International Agricultural Research for Project 8313.

REFERENCES

Putro, S. (1986). Infofish Mark. Dig. **1**, 33.

Wuttijumnong, P. (1987). Studies on Moisture Sorption Isotherms, Salting Kinetics and Drying Behaviour of Fish. PhD Thesis. University of New South Wales, Kensington, NSW, Australia.

Zagarramurdi, A. & Lupin, H.M. (1977). Lat. Am. J. Chem. Eng. Appl. Chem. **7**, 25.

Zagarramurdi, A. & Lupin, H.M. (1980). J. Food Sci . **45**, 1305.

DRYING AND STORAGE OF TROPICAL FISH - A MODEL FOR THE PREDICTION OF MICROBIAL SPOILAGE

P.E. DOE
The University of Tasmania, GPO Box 252C, Hobart, Tasmania, Australia 7001

AND

ENDANG SRI HERUWATI
Research Institute for Fish Technology, PO Box 30, Palmerah, Jarkarta Pusat, Indonesia

INTRODUCTION

Drying is a traditional and widely used method of preserving fish in tropical countries (Waterman, 1976). With adverse climatic conditions or poor preparation and processing, losses of fish due to microbial spoilage, insect infestation and other means can be high (James, 1984). The temperature of the fish during drying and storage together with its water activity (a_w) are the principal factors in determining whether or not the fish will spoil through microbial, enzyme or chemical action (Troller & Christian, 1978).

A model has been developed that predicts microbial spoilage of tropical fish. This model is based on measurements of the growth rates of various bacteria and moulds found on salted and dried fish in Indonesia together with a computer model which calculates the temperature and water activity of the fish during drying and subsequent storage.

MICROBIAL GROWTH RATES

As part of a three year cooperative research programme of the Australian Centre for International Agricultural Research and the Agency for Agricultural Research and Development, Indonesia, a series of measurements was made of the growth rates of various bacteria and moulds isolated from Indonesian dried fish.

Growth rates of two bacteria, <u>Staphylococcus</u> <u>xylosus</u> and <u>Halobacterium</u> <u>salinarum</u>, were measured using a temperature gradient incubator in media with different salt concentrations for control of water activity (McMeekin <u>et</u> <u>al</u>., 1987). Growth rates of the two bacteria were found to vary with temperature and water activity.

Germination times and radial growth rates of 13 moulds found on Indonesian dried fish were measured at 20°, 25°, 30°, 34° and 37°C over a range of water activities from 0.996 to 0.75 on both salt- and sugar-based media (Pitt & Hocking, 1985).

The growth data for the bacteria and moulds have been combined into a single three-dimensional figure (Fig. 1), which shows the time to a particular level of microbial growth at different temperatures and water activities. In the case of bacteria the time is that for an increase in bacteria numbers by a factor of 10^5. For moulds the time is that for germination and growth to a 2 mm diameter colony. The significance of this figure, the "spoilage envelope", is that fish held at a particular temperature and water activity for longer than the time specified is likely to have spoiled.

A complication arises in the case of fish spoiling during drying because the temperature and water activity of the fish are changing as the fish dries. To take this into account a computer model of the drying process has been developed which calculates the temperature and water activity of the fish as it dries.

COMPUTER MODEL OF FISH DRYING

The computer model of fish drying follows that of Jason (1958). Drying occurs in two states, a constant rate period in which convection processes govern the drying rate followed by a period of falling rate when the internal diffusion processes are of prime importance. The computer model relies on the finite difference solution of the fundamental equations governing the diffusion and evaporation of water from fish. Included in the model is a function for calculating the water activity corresponding to a given salt and moisture content using a correlation of the isohalic sorption isotherm of salted fish (Doe *et al*., 1982; Lupin, 1986).

The computer model provides for a solar energy input during drying and thus applies to the traditional sun-drying process as well as drying by mechanical means. It also applies to storage of dried fish - depending on the conditions under which the fish are stored, the fish will desorb or absorb water according to the same processes which occur during drying.

The computer model has been extended to apply to mechanical drying with the experimental data of Wittijumnong (1987); these data were obtained from a series of measurements of the drying behaviour of lemuru (*Sardinella longiceps*) landed at Munchar, East Java and Australian sardine (*Sardinops neopilchardus*) landed in Sydney over a range of temperature from 35° to 55°C, air humidity from 40 to 60%, air velocity from 0.5 to 2.5 m/s and salt content from 0.15 to 0.35 g/g salt-free, fat-free dry basis. A two-term model for diffusion governing drying gave a good fit to the measurements.

The computer model of fish drying has as its starting point the properties of the fish (salt and moisture contents, mass, size, initial microbial load, etc.) together with the external conditions of air temperature, humidity, air speed and solar energy input in the case of sun-drying. As well, certain parameters associated with the heat and mass transfer processes taking place need to be specified, namely the surface heat and mass transfer coefficients and the diffusivity of the water within the fish. These parameters are difficult to predict with any accuracy; they are determined by "fitting" the computer model to measured drying behaviour. This process relies on a set of measurements of drying behaviour made under conditions similar to that for which the model is to be used. A model so determined should be applied only within the limits of operation and for the fish species for which it has been developed and verified.

The computer model calculates the fish temperature and water activity at successive intervals of time. These values are used to determine the relevant bacterial growth rates. The increase in bacterial number is then calculated for the time interval used.

A complication arises from the way in which different micro-organisms have different preferences for certain ranges of temperature and water activities. For example, as the water activity drops during drying below about 0.95, the bacteria present on the wet fish at the start of drying will progressively be overtaken by species more tolerant of high salt concentrations and low water activity. This necessitates restarting the count of bacterial numbers from time to time as the water activity drops.

In calculating the growth of moulds it must be assumed that the fish water activity and temperature are essentially constant; time to spoilage during storage of fish under constant or slowly changing ambient conditions is found from the spoilage envelope for moulds.

The computer programme, "DRYFISH", incorporating the model of fish drying and storage and the spoilage envelope, was written in the PASCAL language to run on IBM-PC type computers. It has been verified for sun-drying of two tropical fish species with different salt contents over a limited range of climatic conditions and for mechanical drying for two medium-fat fish species over a range of values of air temperature, velocity, humidity and salt contents.

MEASURED DRYING AND SPOILAGE BEHAVIOUR

During March 1987 a series of four sun-drying trials was completed in Jakarta, details of which are given in Table 1 (Doe & Heruwati, 1987).

The first trial used chub mackerel (Rastrelliger neglectus) with an average mass of 96 g dry-salted for 21 hours and 45 hours prior to drying. The fish were gutted but not split. A complete set of chemical, bacteriological and organoleptic surveys were carried out on the fish before and after salting and drying. Chemical analysis was for moisture, salt, ash, protein and fat contents; pH, TVB-N, TMA-N and NH_3 were also measured. The bacteriological assessment was a total plate count on Nutrient Agar for fresh fish and Moderate Halophilic Agar (8% NaCl) for fresh and salted fish. Fresh fish was rated organoleptically using a 0-39 point demerit score sheet (Branch & Vail, 1985; Anggawati, 1987) whilst dried fish was assessed on a 0-9 scale with a 5-point scale for assessment of the cooked dried fish. Fish were dried for two days over a three day period during which time the fish mass and temperature were measured at regular intervals and measurements were taken of the air temperature and humidity, wind speed, and cloud cover.

The second trial used slightly larger fish of the same species (average mass 110 g) which were gutted and salted in three separate treatments for 22 hours in 20%, 30% and 40% w/v NaCl brine. The fish were sun-dried for two days.

In order to get a batch of fish to spoil, a third trial was carried out with unsalted mackerel. Fish were gutted, then divided into four lots - half the fish were washed in 3% w/v salt solution after gutting, the other half were unwashed. Half the washed and unwashed fish were sun-dried for one day while the other half were stored inside under ambient conditions. All fish spoiled within 24 hours.

The fourth trial used skipjack (Katsuwonus pelamis), average mass 504 g, gutted and split, then salted for 24 hours by dry-salting, immersion in saturated brine, and immersion in 30% and 20% w/v brine. A fifth treatment was not salted but kept frozen overnight to be dried with the other treatments. All treatments were sun-dried for four days. The unbrined treatment had spoiled by noon on the second day of drying.

Chemical, bacteriological and organoleptic assessments and measurement of drying conditions and behaviour were made for the second, third and fourth trials as for the first trial.

RESULTS OF DRYING RUNS

The results of the drying trials are summarised in Tables 2-5. Drying conditions were far from ideal as indicated by the extent of cloud cover during the trials. However, this suited the purpose of the experiment as in two of the trials, fish spoiled during drying.

Measured bacterial numbers in Trial 4 (Table 5) can be explained by reference to the spoilage envelope (Fig. 1). Bacterial counts on the unbrined and 20% brined fish increased by more than three orders of magnitude ($2.0x10^4$ to $8x10^7$ and $4.3x10^4$ to $7.5x10^7$, respectively) while bacterial counts from the more heavily salted fish increased by around a factor of 10 only. All of the drying of the unbrined and 20% brined fish was at a water activity in excess of 0.86 where Fig. 1 shows that S. xylosus predominates and grows rapidly. However, the water activity of the more heavily salted fish is already below 0.86 before drying starts and thus the more slowly growing Halobacterium sp. predominate resulting in lower bacterial counts from the dried fish. A similar trend is evident in trial 2 (Table 3) which shows that the bacterial count for the 20% brined fish increased from $2x10^4$ to $1.2x10^7$ during drying while that in the fish salted in 40% brine increased from $4.3x10^3$ to $1.3x10^5$.

TABLE 1
Details of sun drying trials, March 1987, Jakarta, Indonesia

Date	10-12 Mar.	17-18 Mar.	19 Mar.	23-27 Mar.
Fish species:	R. neglectus	R. neglectus	R. neglectus	K. pelamis
Trial Number:	1.	2.	3.	4.
Fish mass (g):				
- whole	114.9	132.1	115.7	504.0
- gutted	95.8	110.1	92.1	449.0
Fish size (mm):				
- length	171*	173*	171*	180**
- thickness	22.6	26.1	22.6	22.8
- width	46	48	46	150
Treatment:	Saturated brine	Three brine concen-trations	Unbrined	Four salted one unsalted

* Length of body not including caudal fin.
** Length of body not including head or tail.

FITTING AND VERIFICATION OF THE COMPUTER DRYING MODEL

Values of the surface heat and mass transfer coefficients and the diffusion coefficient were chosen so that the computer model gave values of fish mass and temperature close to those measured during the first sun-drying trial, using the measured fish dimensions and average values of the measured climatic conditions. The results are shown in Fig. 2. The same values of heat and mass transfer coefficients together with the measured fish dimensions and climatic data gave credible agreement with the measured drying behaviour for the subsequent three trials as shown in Figs. 3,4 and 5. It was, however, necessary to adjust the value of the diffusion coefficient by 20% to allow for different salt contents.

As Fig. 1 illustrates, different bacteria predominate at different temperatures and water activities; <u>Staphylococcus</u> <u>xylosus</u> is the predominant bacterium over a wide range of water activities above about 0.85. Below this value, <u>Halobacterium</u> <u>salinarum</u> and moulds grow at a faster rate. This behaviour was modelled by calculating the growth of both organisms and choosing the greater number. When the water activity drops below the point of zero growth rate of <u>S.xylosus</u>, the number of bacteria is held constant until it is exceeded by the numbers of <u>H.salinarum</u>.

TABLE 2
Results of drying trial 1, 10 - 12 March 1987, Jakarta

Organoleptic: Fresh fish demerit 19 (39 max demerit)
Dried fish (brined 21 hours) 8.9 (9 max score)
Dried fish (brined 45 hours) 8.7 (9 max score)

Bacteria (Total plate count/g):

	Nutrient Agar	MHA	(a_w)
Fresh fish	$2.2x10^5$		.998
Fish brined 21 hr		< 300	.850
Fish brined 21 hr and dried	$1.2x10^4$	$2.3x10^3$	.736
Fish brined 45 hr		< 300	.794
Fish brined 45 hr and dried	< 300	< 300	.737

Chemical:

	Moisture cont. % wet basis	Salt cont. % dry basis*	Fat % w.b.	pH
Before brining	77.9	1.4	2.5	6.2
Fish brined 21 hours	58.4	39.7	1.5	6.1
dried 24 hours	50.8	36.2		
Fish brined 45 hours	57.2	39.3	0.9	6.0
dried 24 hours	49.8	35.2		

Drying conditions:

	Day 1 mean (min-max)	Day 2 mean (min-max)	Day 3 mean (min-max)
Air temp. (°C)	32.4 (32 - 33)	31.0 (29 - 32)	32.4 (31 - 34)
Air rel.hum. (%)	61.9 (57 - 66)	68.0 (60 - 83)	61.0 (54 - 70)
Wind speed (m/s)	1.2 (.79-1.6)	1.2 (.78-2.1)	1.1 (1.2-2.0)
Cloud (oktas)	3 (1 - 6)	5 (2 - 8)	5 (2 - 8)

Drying behaviour (average mass/fish in g):

	Fish brined 21 hours	Fish brined 45 hours
Number of fish	33	39
Mass before brining	95.8	95.8
Mass after brining	87.9	86.7
Mass after day 1	75.5	-
Mass after day 2	68.5	76.2
Mass after day 3	-	73.1

* Mass of salt/mass of moisture free solids (%)

TABLE 3
Results of drying trial 2, 17 - 18 March 1987, Jakarta

Organoleptic:

Fresh fish demerit	22.8
Dried fish (40% brine) score	7.7
Dried fish (30% brine) score	8.0
Dried fish (20% brine) score	6.7

Bacteria (Total plate count/g):

	Nutrient Agar	MHA	(a_w)
Fresh fish	$5.2X10^5$		.997
After 21 hrs in 20% brine	$2.0X10^4$	$3.3X10^3$	.940
Dried 2 days (20% brine)	$1.2X10^7$	$5.3X10^4$	.862
After 21 hrs in 30% brine	$6.5X10^3$	$5.0X10^3$	.880
Dried 2 days (30% brine)	$6.7X10^6$	$8.0X10^3$	.821
After 21 hrs in 40% brine	$4.3X10^3$	$4.1X10^3$	.834
Dried 2 days (40% brine)	$1.3X10^5$	$2.5X10^3$	.728

Chemical:

	Moisture cont. % wet basis	Salt cont. % dry basis	Fat % w.b.	pH
Before brining	77.8	4.8	0.6	6.0
After 20% brine	70.1	25.2	1.5	6.1
dried 2 days	59.0	33.4		
After 30% brine	68.7	27.8	0.1	6.1
dried 2 days	56.2	32.5		
After 40% brine	64.6	31.9	0.7	6.1
dried 2 days	55.4	35.6		

Drying conditions:

	Day1 mean (min-max)	Day2 mean (min-max)
Air temp. (^oC)	32.2 (31 - 33)	32.2 (29 - 34)
Air rel.hum. (%)	62.1 (58 - 69)	62.6 (52 - 79)
Wind speed (m/s)	.68 (0.1-3.6)	.43 (0.1-0.8)
Cloud (oktas)	5 (1 - 8)	3 (0 - 6)

Drying behaviour (average mass/fish in g):

	Fish in 20% brine	30% brine	40% brine
Number of fish	21	23	23
Mass before brining	110.1	110.1	110.1
Mass after brining	111.1	106.7	107.6
Mass after day 1	90.0	87.8	93.3
Mass after day 2	76.1	76.5	82.9

TABLE 4
Results of drying trial 3, 19 March 1987, Jakarta

Organoleptic: Fresh fish demerit 26
Spoiled fish demerit 35

Bacteria (Total plate count/g):

	Nutrient Agar	(a_w)
Fresh fish	$1.7X10^7$	.985
Washed then kept indoors	$7.7X10^6$	
Not washed then kept indoors	$4.5X10^6$	
Washed then dried	$1.8X10^7$	
Not washed then dried	$1.5X10^7$	

Chemical:

	Moisture cont. % wet basis	Salt cont. % dry basis	Fat % w.b.	pH
Fish before treatment	76.3	2.2	1.2	6.1

Drying conditions:

	Outdoors mean (min-max)	Indoors mean (min-max)
Air temp. (^{o}C)	32.6 (31 - 33)	30.2 (29 - 32)
Air rel.hum. (%)	62.5 (60 - 69)	67.0 (61 - 69)
Wind speed (m/s)	.28 (0 -1.0)	-
Cloud (oktas)	5 (4 - 7)	-

Drying behaviour (average mass/fish in g):

	Outdoors		Indoors	
	unwashed	washed	unwashed	washed
Number of fish	19	18	18	18
Mass before drying	103.2	94.4	95.0	109.0
Mass after 5.5 hrs	89.0	80.6	91.1	95.0
Mass after 12.5 hrs	-	-	85.0*	88.9*
Mass after 22.3 hrs^{+}	82.1*	75.0*	-	-

* Fish assessed to be spoiled.

\+ Fish held at ambient temperature (29°C) overnight had spoiled by morning.

TABLE 5

Results of drying trial 4, 23 - 26 March 1987, Jakarta

Organoleptic:	Fresh fish demerit	9.5
	Dried fish scores-dry salted	6.8
	- saturated brine	7.4
	- 30% brine	7.4
	- 20% brine	7.6
	- unsalted	5.5

Bacteria (Total plate count/g):

	Nutrient Agar	MHA	(a_w)
Fresh before treatment	$4.5X10^4$		
Fish frozen overnight	$2.0X10^4$		.994
dried (unbrined)	$8.0X10^7$		1.000
Fish after 20% brine	$4.3X10^4$	$2.2X10^4$	.922
dried (20% brine)	$7.5X10^7$	$3.1X10^7$	.864
Fish after 30% brine	$6.2X10^3$	$1.2X10^3$	.855
dried (30% brine)	$1.7X10^4$	$1.2X10^4$	.810
Fish after sat. brine	$3.6X10^3$	$2.7X10^3$	.866
dried (sat. brine)	$6.7X10^4$	$3.8X10^4$	.767
Fish after dry salting	$1.2X10^5$	$4.2X10^4$	.772
dried (dry salted)	$3.5X10^5$	$3.2X10^5$	.770

Chemical:

	Moisture cont. % wet basis	Salt cont. % dry basis	Fat % w.b.	pH
Fish before treatment	74.2	4.7	1.2	6.2
dried 4 days	65.2	3.2		6.8
After 20% brine	68.6	29.1	0.7	6.2
dried 4 days	55.2	26.4		6.0
After 30% brine	63.9	27.2	1.1	6.3
dried 4 days	54.5	31.5		6.0
After sat. brine	62.1	37.2	1.8	6.0
dried 4 days	53.2	42.3		5.9
After dry salting	-	32.0	0.5	6.1
dried 4 days	50.5	37.1		6.0

Drying conditions:	Day1	Day2	Day3	Day4
Air temp. (^{o}C)	30.5	30.7	30.7	32.4
Air rel.hum. (%)	70.2	69.8	71.5	65.0
Wind speed (m/s)	0.2	0.2	0.7	1.2
Cloud (oktas)	3	3	3	2

Drying behaviour (average mass/fish in g):

	Dry salted	Sat. brine	30% brine	20% brine	unbrined
Number of fish	2	2	2	2	2
Mass before brining	468	460	500	458	468
Mass after brining	383	406	468	437	450
Mass after day 1	344	357	413	382	388
Mass after day 2	313	319	366	328	316
Mass after day 3	300	303	348	307	280
Mass after day 4	294	294	336	287	248

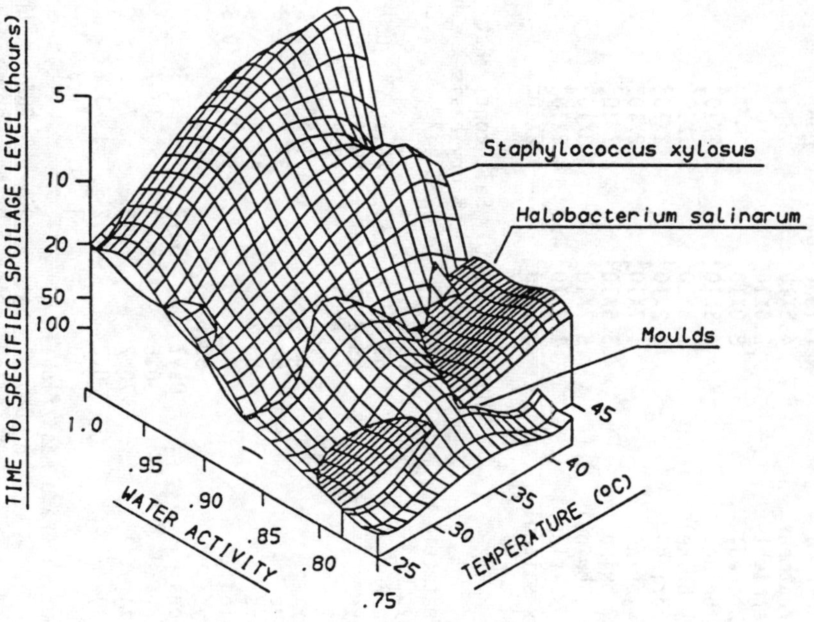

Figure 1. Spoilage envelope- times to spoilage based on measured growth characteristics of certain moulds and bacteria.

Bacterial numbers at the end of drying were found to be significantly less than those calculated from unadjusted bacterial growth rates measured in the laboratory. This may be due to limitations in the rate at which bacteria diffuse into the fish muscle. (The laboratory measurements were conducted in a shaken liquid medium). There is also the inhibiting effect of direct sunglight on bacterial growth (McCambridge & McMeekin, 1981). In order to get reasonable agreement between computed and measured bacterial numbers at the end of drying it was necessary to increase the generation times for bacteria during the period of sun-drying by a factor of 4.6. This factor was arrived at by trial and error so that bacterial numbers exceeded 10^8 at around midday on the second day of drying the unbrined sample in trial 4 (see below), at which time the fish was adjudged organoleptically to have spoiled.

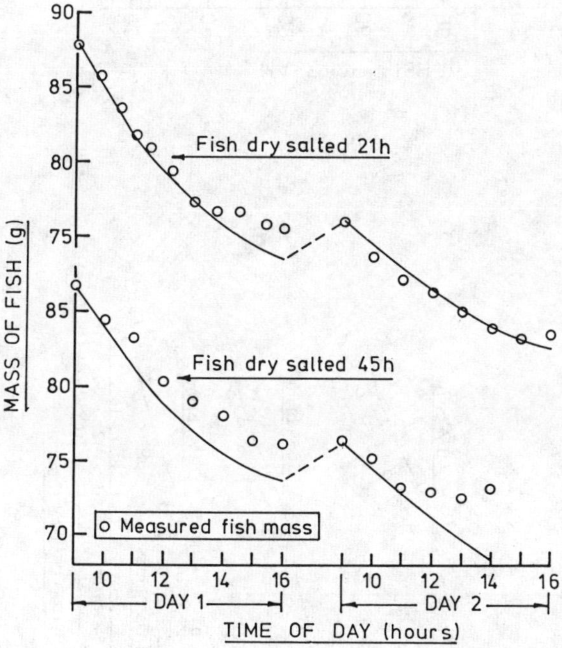

Figure 2. Measured and computed drying behaviour of mackerel (R. neglectus) - Trial 1.

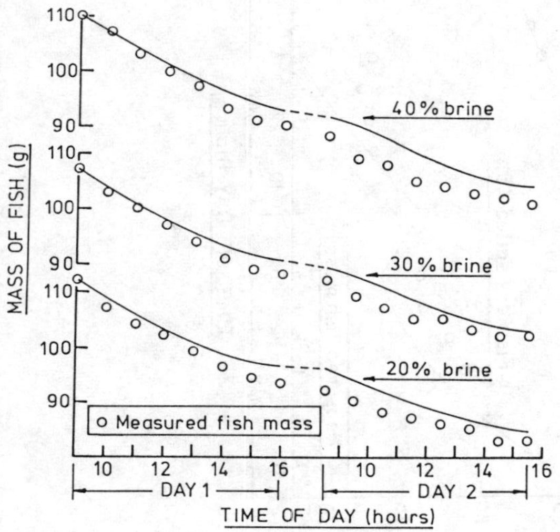

Figure 3. Measured and computed drying behaviour of mackerel (R. neglectus) - Trial 2.

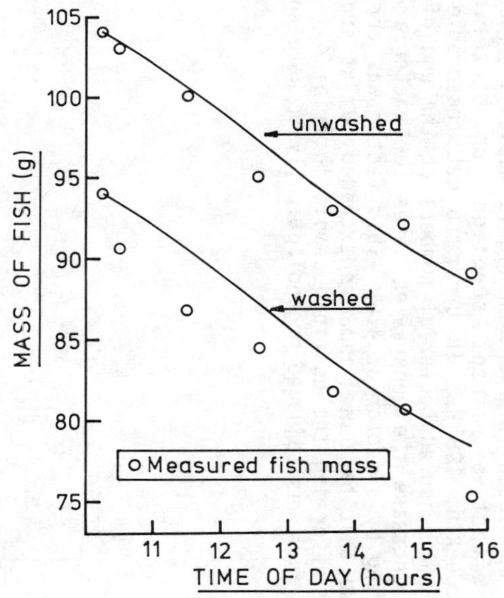

Figure 4. Measured and computed drying behaviour of mackerel (R. neglectus) - Trial 3.

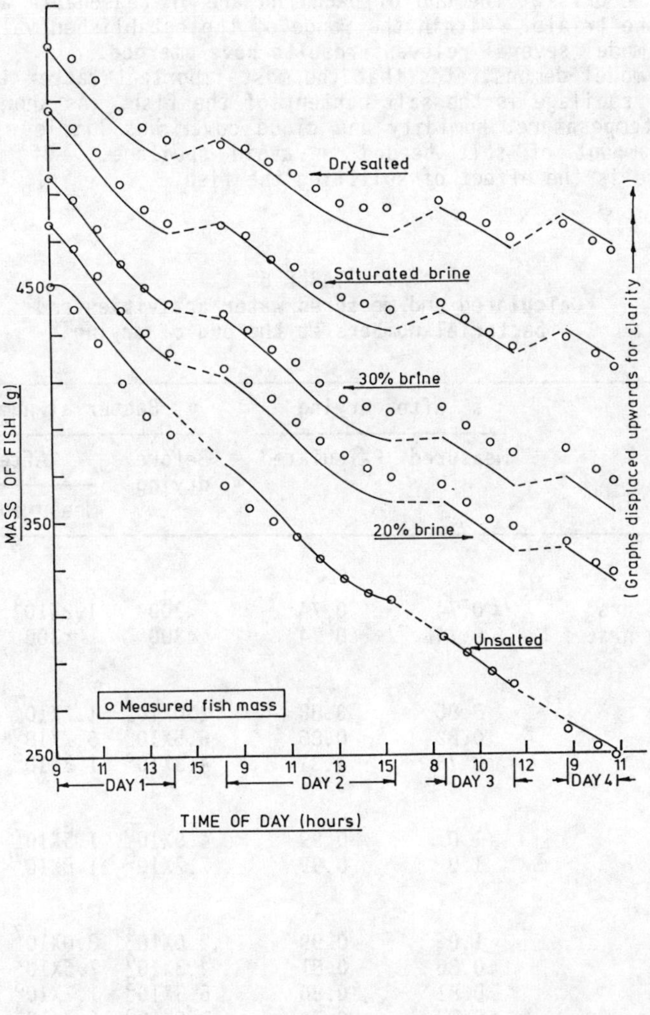

Figure 5. Measured and computed drying behaviour of skipjack (K. pelamis) - Trial 4.

RESULTS FROM THE COMPUTER MODEL

Spoilage During Drying

Fig. 6 shows measured and computed temperature of one of the treatment in the second trial (30% brine), together with the calculated values of water activity and bacterial numbers.

Table 6 shows that the calculated and measured water activities and bacterial counts at the end of each run are in reasonable agreement for most of the trials. Within the range of the established validity of the computer model several relevant results have emerged.

The model demonstrates that the most important factor in preventing bacterial spoilage is the salt content of the fish. As shown in Fig. 7, the air temperature, humidity and cloud cover has little effect on the minimum amount of salt needed to avoid spoilage. Of much greater importance is the effect of splitting the fish.

TABLE 6
Calculated and measured water activities and bacterial numbers at the end of drying

	a_w after drying		Bacterial numbers		
	Measured	Calculated	Before drying	After drying	
				Measured	Calculated
Trial 1					
Brined 21 hrs	0.74	0.74	<300	$1.2X10^4$	$6.8X10^2$
Brined 45 hrs	0.74	0.74	<300	<300	$6.5X10^2$
Trial 2					
20% brine	0.86	0.88	$2.0X10^4$	$1.2X10^7$	$3.3X10^8$
30% brine	0.82	0.86	$6.5X10^3$	$6.7X10^6$	$7.4X10^6$
40% brine	0.73	0.81	$4.3X10^3$	$1.3X10^5$	$2.3X10^3$
Trial 3					
Unwashed	1.0	0.99	$4.5X10^6$	$1.5X10^7$	$3.8X10^7$
Washed	1.0	0.99	$7.7X10^6$	$1.8X10^7$	$6.6X10^7$
Trial 4					
Unbrined	1.0	0.99	$2.0X10^4$	$8.0X10^7$	$> 10^{10}$
20% brine	0.86	0.81	$4.3X10^4$	$7.5X10^7$	$9.3X10^6$
30% brine	0.81	0.80	$6.3X10^3$	$1.7X10^3$	$2.6X10^5$
Sat. brine	0.77	0.74	$3.6X10^3$	$6.7X10^4$	$1.5X10^4$
Dry salted	0.77	0.74	$1.2X10^5$	$3.5X10^4$	$4.4X10^3$

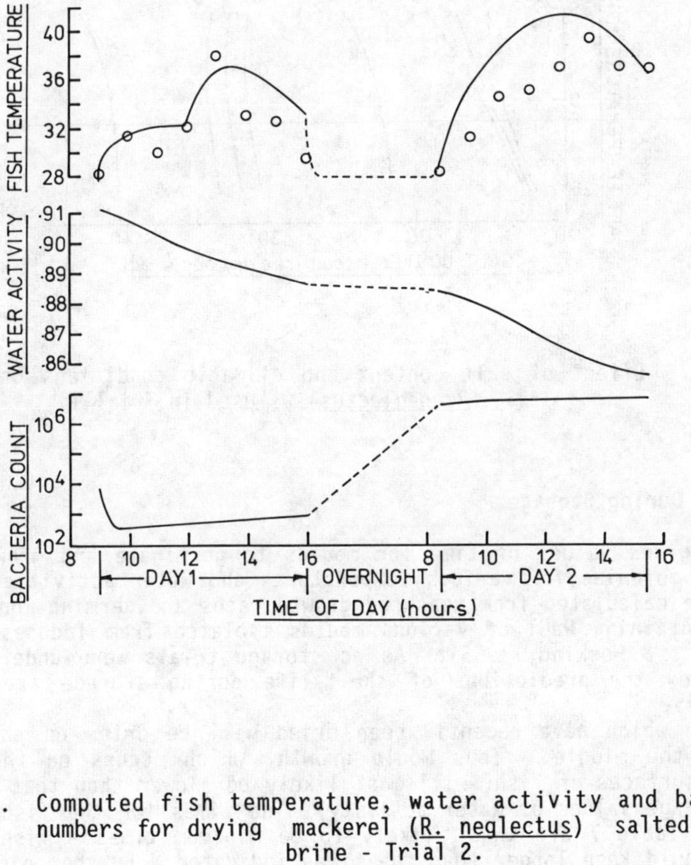

Figure 6. Computed fish temperature, water activity and bacterial numbers for drying mackerel (R. neglectus) salted in 30% brine - Trial 2.

With the conditions encountered in the trials, the lightly salted fish lost weight more rapidly than more heavily salted fish (cf. Jason & Peters, 1973). Also, the fish with low salt content lost weight during the night while the more heavily salted fish gained weight in some cases. So far as the likelihood of spoilage was concerned, any slowing down of drying due to increased salt content is more than offset by the effect of salt in rapidly lowering the water activity. Computed values of water activity for the fish salted for 24 hours in saturated brine in the fourth sun-drying trial had dropped to 0.80 after the first day's drying whereas the unsalted fish still had a water activity close to 1.0 after four days.

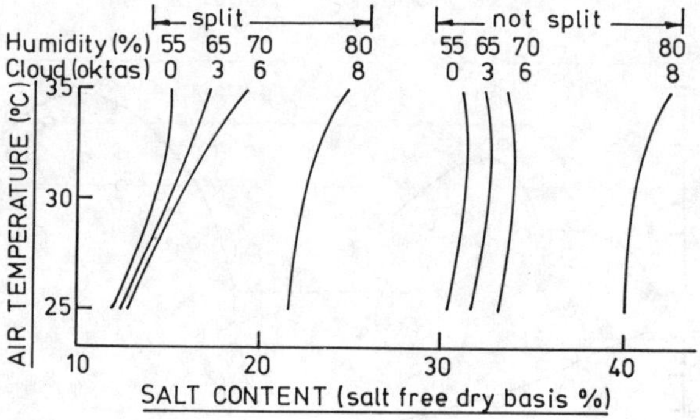

Figure 7. Effect of salt content and climatic conditions on mackerel (R. neglectus) as used in Trial 1.

Spoilage During Storage

Table 7 gives values of time for moulds to germinate and grow to 2 mm diameter colonies for various temperatures and water activities. These data were calculated from measured growth rates and germination times on media containing NaCl of various moulds isolated from Indonesian dried fish (Pitt & Hocking, 1985). As no storage trials were undertaken for this study the predictions of shelf-life during storage are entirely speculative.

Fish which have recently been dried will be drier on the outside than in the middle. Thus mould growth, which occurs on the exposed outside surfaces of fish, will most likely be slower than that predicted from average values of water activity. The times for the fish to spoil given in Table 7 are thus likely to be minimum times - fish recently dried should keep longer than the times indicated. Further experimental verification is required before the model of storage can be extended to situations where store conditions vary with time.

APPLICATION OF THE COMPUTER MODEL

The question might be asked - what is the use of modelling such a complicated and variable process as fish drying and storage? One answer is that because the process is so variable it is difficult and expensive to undertake measurements of spoilage under conditions normally encountered in traditional sun-drying of fish in tropical countries; there is no way that climatic conditions can be varied to suit the

TABLE 7
Time (in days) for moulds to germinate and grow to 2 mm diameter at various temperatures and water activities (Pitt & Wheeler, unpublished).

	Temperature (^{o}C)		
a_w	25	30	35
0.99	0.75	0.50	0.45
0.90	1.4	1.0	1.3
0.85	3.7	2.0	3.3
0.80	11.3	3.4	10.1
0.75	29	14	28

experimenter. The computer simulation, if set up to give a reasonable fit over a range of climatic conditions, can be used with some confidence to predict the effects of changes in the various variables involved.

Another use for the model is to evaluate the limits of fish size, salt content, thickness, etc. which will allow the fish to be dried under a given set of climatic conditions without spoiling.

The model is also useful in giving the user a "feel" for the effects of the different variables - for example, it demonstrates the effects on the drying rate and growth rates of bacteria of splitting the fish. The model can be used to determine when climatic conditions, or fish size, or limitations on salt content would necessitate the use of some form of solar or mechanical dryer.

CONCLUSIONS

This work is an attempt to apply to the fish drying industry some of the computer modelling techniques which have proven successful in other industries. Personal computers are now commonplace in research institutions world-wide. Research and extension personnel can use a model of fish drying and storage to suggest improvements to traditional practices. Computer simulation is a very effective method of learning about a process - it is easier to experiment with a computer model than to assimilate the theory or gain the practical experience on which the model is based.

It is to be hoped that further development of the model will occur and that its use may eventually extend to the fish processors.

ACKNOWLEDGEMENTS

This work has resulted from a three year cooperative research programme supported by the Agency for Agricultural Research and Development, Indonesia and the Australian Centre for International Agricultural Research. The authors would like to record their appreciation for the assistance rendered by Dr. J.N. Olley, Dr. J.I. Pitt, and Ms. K.A. Wheeler, CSIRO; Dr. T.A. McMeekin, Dr. J.L. Madden and Mr. R. Chandler, University of Tasmania; Ms. E. Gorczyca and Dr. J. Sumner, Royal Melbourne Institute of Technology; and Dr. Sumpeno Putro, Miss Agnes Anggawati, Mrs. Ninoek Indriati, Mrs. Miryam Asatyasih, Miss Murniyati and Miss Nunuk of the Research Institute for Fish Technology, Jakarta. The authors are particularly grateful for permission to incorporate results from Dr. Wuttijumnong's PhD thesis into the drying model.

REFERENCES

Anggawati, A.M. (1987). Mesophilic spoilage of milkfish (Chanos chanos). (Submitted for publication - Asean Food Journal).

Branch, A.C. & Vail, A.M.A. (1985). Food Technol. Australia **37**(8), 352.

Doe, P.E., Rahila Hashmi, Poulter, R.G. & Olley, J. (1982). J. Food Technol. **17**, 125.

Doe, P.E. & Endang Sri Heruwati. (1987) A model for the prediction of the bacterial spoilage of sundried tropical fish. (Submitted for publication - J. Food Eng.

James, D. (1984). Infofish Market. Digest. **4**, 41.

Jason, A.C. (1958). Fundamental Aspects of the Dehydration of Foodstuffs, p. 103-135. Society of Chemical Industry, London.

Jason, A.C. & Peters, G.R. (1973). J. Phys. D. : Appl. Phys. **6**, 512.

Lupin, H.A. (1986). Cured Fish Production in the Tropics. (Reilly, A. & Barile, L.E., eds), p. 16. Proceedings of a Workshop on the production of cured fish. University of the Philippines in the Visayas, Diliman, Quezon City, Philippines, April 1986.

McCambridge, J. & McMeekin, T.A. (1981). Appl. Environ. Microbiol. **41**(5), 1083.

McMeekin, T.A., Chandler, R.E., Doe, P.E., Garland, C.D., Olley, J., Putro, S. & Ratkowsky, D.A. (1987). Model for combined effect of temperature, salt concentration/water activity on the growth rate of Staphylococcus xylosus. J. appl. Bacteriol. (in press)

Pitt, J.I. & Hocking, A.D. (1985). Mycotaxon **22**, 197.

Troller, J.A. & Christian, J.H.B. (eds) (1978). Water Activity and Food. Academic Press, New York.

Waterman, J.J. (1976). The production of dried fish. FAO Fish. Tech. Pap., No. 16.

Wuttijumnong, P. (1987). Studies on moisture sorption isotherms, salting kinetics and drying behaviour of fish. PhD thesis, University of New South Wales.

STABILITY OF *DENDENG*

K.A. BUCKLE, H. PURNOMO AND S. SASTRODIANTORO
Department of Food Science and Technology
The University of New South Wales
P.O. Box 1, Kensington, NSW 2033
Australia

INTRODUCTION

Tropical conditions provide a severe storage environment for any food, including dehydrated and intermediate moisture (IM) foods. Preservation of foods by moisture removal with or without solute uptake is relatively simple and inexpensive and is consequently widely practised in tropical regions. Nevertheless, high storage temperatures and humidities and high initial levels of microbial contamination can lead to significant levels and rates of spoilage from chemical, physical and microbiological changes unless production and storage conditions are controlled.

Dendeng is not a staple food, but is widely consumed in many regions of Indonesia as a snack or as an accompaniment to a range of dishes. Before consumption it is deep-fried for a short time. It can be consumed soon after production but is found increasingly in markets in plastic or other packaging.

PRODUCTION AND COMPOSITION OF *DENDENG*

Production

Production of *dendeng* in Indonesia generally is limited in scale to what can be made in the home or in butcher's shops, with relatively few large-scale processors. Production statistics for *dendeng* in Indonesia are not available, but substantial quantities are produced.

Dendeng is a traditional Indonesian IM meat product containing coconut sugar, salt and spices (Purnomo *et al*., 1983). It is prepared from meat in the form of thin slices (*dendeng sayat*) or from minced meat (*dendeng giling*). It may be prepared from beef, chicken, pork or fish, but dried beef (*dendeng sapi*) is the major product found in the markets. *Dendeng* has a sweet taste due to its high sugar content, and together with the strong flavour of the spices and the dried meat gives *dendeng* a characteristic flavour that differentiates it from other traditional IM meats, e.g. jerky, biltong, etc.

A typical flow chart for *dendeng* production is shown in Fig. 1. Fresh meat is sliced to about 3 mm thick and soaked for up to 12 h at ambient temperature in a mixture of sugar, salt and spices (Table 1) added to a small volume of water and then spread on bamboo trays and sun-dried. The sugar used is coconut or palm sugar, commonly produced by a cottage industry from the coconut palm (*Cocos nucifera*) primarily, or from other palms, by collecting the juice or toddy from the palm tree after cutting the spathes forming the inflorescence, concentrating the juice by heating and then crystallising the sucrose to form a brownish crystal (*gula kelapa*) containing a variety of coloured impurities. It is moulded in a half coconut shell.

The spices used commonly are coriander, tamarind, garlic and the root of the greater galangal, but some recipes may also contain onion, pepper and other spices. Coriander is the fruit of *Coriandrum sativum* L; the dried, nearly ripe fruits (coriander seeds), or fresh and dried seeds, are used. The fruit contains 0.4-1% volatile oil containing about 65% d-linalool (coriandrol), as well as 26% fat, 17% protein, 10% starch and 20% sugars. It possesses strong lipolytic activity as well as some antioxidative activity. Tamarind is a sour tropical fruit of the large tree, *Tamarindus indica*, containing 8-12% tartaric acid and 21-30% sugars. Tamarind pulp is used as an acidifying agent in many traditional recipes. Garlic is the compound bulb of *Allium sativum* L. Roots of greater galangal are the dried rhizomes of *Kempferia galangal* or the greater galangal, and contain 0.5-1% essential oil with a spicy, ginger-like odour. When available, the fresh material is preferred.

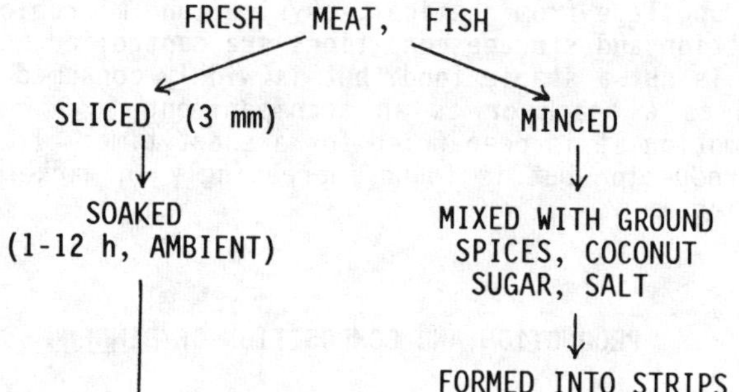

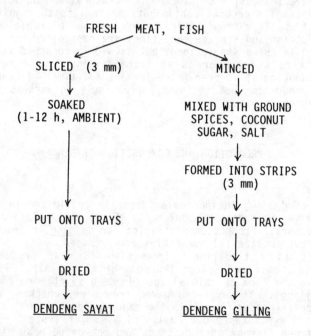

Figure 1. Flow chart for *dendeng* production.

There is no preferred recipe for dendeng production, as slightly different recipes are found from one region to another. The major variations are the proportions of the non-meat ingredients and the variety of spices added.

For dendeng giling, meat is minced and mixed with the dried, ground spices, sugar and salt4, and then rolled into a thin sheet about 3 mm thick, placed onto bamboo trays, and sun-dried.

For experimental purposes at UNSW, dendeng giling was prepared from round steak or mullet fillets using a modified traditional recipe (Table 1) containing a higher proportion of non-meat ingredients. Coconut sugar was obtained from three areas in Indonesia (Malang, Yogyakarta and Denpasar), while all ot4her dried spices were from Asian food shops in Sydney, Australia. After blending, the mixture was rolled to a thickness of about 3 mm and dried for 50 min at 70°C in a cross-draught dehydrator.

TABLE 1

Meat and fish dendeng giling recipes

Component	Weight (g)	
	Typical traditional	Experimental
Minced meat, fish	1000	1000
Coconut sugar	200	308
Coriander	60	121
Cooking salt	55	55
Tamarind	8	12.4
Garlic	7	10.8
Roots of Galangal	2	3

There are significant variations in traditional fish dendeng production. While some villagers soak fish fillets in a marinading solution as for meat dendeng, some soak the fillets for 5-10 min in a saturated salt brine, then pour coconut sugar on the surface and leave it for up to 20 h. The fish is drained, coriander (and perhaps other spices) and some sugar are sprinkled on the surface, and the fish then sun-dried for 8 h. Minced fish dendeng is also produced traditionally by a method similar to that for meat dendeng giling.

Composition

A typical proximate composition of dendeng is shown in Table 2. Meat dendeng giling produced in the laboratory typically has a moisture content of about 25% (dry basis), sugar content about 35%, protein content about 35%, fat content about 10%, salt content 7-8%, pH 5.5, a_w about 0.65, with relatively low levels of microorganisms. The fish dendeng we have produced has a slightly higher moisture content, lower fat and protein contents, and a similar sugar content. We have not been able to analyse at UNSW any traditional Asian dendeng samples because of our strict quarantine regulations, but analyses in Indonesia show quite a variable moisture content, with some sliced dendeng samples being either pliable or quite tough. Microbiologically these products are safe and do not spoil unless produced at significantly higher a_w, e.g. >0.8-0.85, or are stored in humid atmospheres.

The coconut sugars used in our work vary in sucrose content between about 70 and 80%, with smaller levels (6-18%) of glucose and fructose (Table 3). The crystallised sugars are generally brownish and contain considerable impurities that contribute to their flavour and hence their appeal in traditional recipes. They are also easily prepared and hence readily available in the market and cheaper than more refined samples.

TABLE 2
Typical chemical composition (% dry basis) of minced meat and fish dendeng

Component	Meat	Fish
Moisture	24.1 - 27.4	32.4 - 35.4
Protein	33.4 - 36.9	31.8 - 32.3
Fat	9.6 - 10.7	3.0 - 3.8
Sugar	33.9 - 35.7	N.A.
Salt	7.4 - 8.4	N.A.
Ash	10.9 - 12.5	N.A.
a_w	0.62- 0.66	0.66- 0.68

N.A. = not analysed

TABLE 3
Composition of coconut sugars (Purnomo, 1986)

Origin	a_w	Moisture content (% dry basis)	Sugars (g/100 g)		
			Sucrose	Glucose	Fructose
Malang	0.69	11.9	70.9	3.0	2.9
Yogyakarta	0.69	11.7	79.0	3.5	3.1
Denpasar	0.63	9.7	70.5	9.0	9.0

SUBSTITUTION OF MEAT IN DENDENG

In many areas of Indonesia meat is either expensive or not readily available. Some of our early studies on dendeng examined locally available and inexpensive substitutes for some of the meat in dendeng giling recipes (Purnomo et al., 1984). Two such substitutes were breadfruit, a very large fruit often cooked as a vegetable, but which contains 3-7% protein (pulp) as well as Fe, Ca, K, riboflavin and niacin, and corn grits, a low grade by-product from corn milling. Substitution of 25% of the meat by 15% breadfruit + 10% corn grits (Table 4) produced a dendeng that was still organoleptically acceptable. Higher levels of substitution gave products that were markedly yellowish or fibrous and lacked the typical dendeng character.

STABILITY OF DENDENG

Dendeng is an IM meat containing significant levels of protein, reducing sugars and fat. During storage at tropical temperatures, it can and does undergo a variety of chemical and physical changes that eventually render the product unacceptable.

Non-Enzymic Browning (NEB)

Significant browning of dendeng can occur during storage, even though the product initially is dark in colour. We have found that erroneously high results can be obtained for browning measurements if the absorbance due to the meat pigments and spices, etc. is not taken into account (Buckle & Purnomo, 1986). NEB initially increases, but then decreases during extended storage (for dehydrated meat also), perhaps as NEB pigments become involved in interactions with oxidising lipids (Table 5).

TABLE 4
Acceptability of dendeng giling containing breadfruit and corn grits (Purnomo et al., 1984)

Minced meat (% w/w)	Bread-fruit (% w/w)	Corn grits (% w/w)	Colour*	Flavour*	Texture*	Overall accepta-bility
100	-	-	7	7	6	7
90	10	-	6	6	6	6
80	20	-	6	5	6	6
70	30	-	5	5	5	5
100	-	-	7	7	7	7
90	-	10	6	5	5	5
80	-	20	4	5	4	5
70	-	30	3	3	4	4
100	-	-	7	7	7	6
75	15	10	6	6	6	6
50	30	20	3	3	4	5
25	45	30	3	3	4	2

* Mean scores for 60 panelists for rankings on 9-point hedonic scale (9 = like extremely, 1 = dislike extremely)

The non-meat ingredients of dendeng increase significantly the level of NEB over that contributed by the meat alone. Coconut sugar increases NEB while salt appears to decrease NEB if spices are present (Table 6).

During storage of dendeng, hydrolysis of sucrose occurs, with glucose reacting in NEB reactions more rapidly than fructose (Table 7). In minced fish dendeng, NEB is higher in unfried than in fried dendeng, and slightly higher in unfried dendeng stored in polyethylene bags than in tinplate containers.

Lipid Oxidation

The chemical changes in the lipids of a product such as dendeng are quite complex - lipids are derived not only from the meat but also from spices such as coriander. Metal impurities in fining agents used to clarify

TABLE 5
NEB of dendeng and dehydrated meat during storage at 50°C (Buckle & Purnomo, 1986)

Sample	Storage time (weeks)	Absorbance at 420 nm	
		Aqueous trypsin digestion*	Lead acetate - alcohol precipitation
Dendeng	0	0.535	0.009
	4	3.145	0.115
	8	4.210	0.159
	12	3.190	0.111
Dehydrated meat	0	0.275	0.005
	4	1.205	0.045
	8	1.250	0.050
	12	1.150	0.047

* Filtrate diluted 1:5 before absorbance measurement

sugar syrups before crystallization may also increase the tendency of fats to undergo oxidation.

The thiobarbituric acid (TBA) method does not appear to be a satisfactory method for measuring the extent of lipid oxidation in meat dendeng, since TBA numbers tend to fluctuate during storage. TBA numbers invariably decrease during the early stages of storage, perhaps as intermediates in lipid oxidation react in NEB reactions, or as NEB products or precursors show some antioxidative activity. Subsequent increases in TBA number during storage (e.g. >6 months at 37°) are not always related to the sensory assessment of rancid odours or flavours.

We have also examined the use of C_1-C_5 hydrocarbons (e.g. methane and pentane) as indicators of autoxidation of dendeng. Pentane has been shown by many workers to increase significantly during the oxidation of linoleic acid from a variety of lipids and lipid-containing foods. From some freeze-dried meats, however, methane was the predominant hydrocarbon. In dendeng stored for up to 6 weeks at 37°C, headspace methane levels decreased significantly, paralleling decreases in TBA numbers, and no pentane was found. Whether the hydrocarbons are produced from the meat or the spices is not known.

TBA numbers initially decreased, then increased when minced fish dendeng was stored at 30°C in plastic bags; in tinplate containers, TBA numbers decreased during storage. TBA numbers increased in dendeng fried before storage while TBA of unfried dendeng decreased.

TABLE 6
Effect of non-meat ingredients on NEB at 50°C
(Buckle & Purnomo, 1986)

Ingredients	Storage time (weeks)	Absorbance at 420 nm: Aqueous trypsin digestion*	Absorbance at 420 nm: Lead acetate - ethanol precip^{n}
Minced meat	0	0.40	0.00
	6	1.37	0.06
Minced meat +	0	0.51	0.02
coconut sugar	6	3.89	0.16
Minced meat +	0	0.52	0.01
coconut sugar + cooking salt	6	3.36	0.14
Minced meat +	0	0.73	0.03
coconut sugar + cooking salt + spices (= dendeng)	6	3.94	0.16
Minced meat +	0	0.82	0.03
coconut sugar + spices	6	3.21	0.13
Minced meat +	0	0.33	0.01
cooking salt	6	1.57	0.06
Minced meat +	0	0.49	0.02
cooking salt + spices	6	2.55	0.10
Minced meat +	0	1.01	0.02
spices	6	2.65	0.11

* Filtrates were diluted 1:5 before absorbance measurement

TABLE 7
Sugar components (% wet weight) of dendeng stored at 50°C (Purnomo, 1986)

Storage time (weeks)	Sucrose	Glucose	Fructose	Xylose
0	18.0	1.6	1.2	Trace
4	5.0	2.4	6.3	0.5
8	4.2	1.8	5.7	0.6
12	1.7	1.5	5.8	0.6

The reactions involved overall are complex; moisture and hence a_w changes in dendeng stored in polyethylene bags, especially in tropical climates, can change the extent of rancidity during storage. However, packaging unfried dendeng in airtight containers would appear to give maximum storage life as far as flavour changes are concerned.

Available Lysine and Protein Solubility

Available lysine is lower in dendeng than in dehydrated meat, and lower when measured by the dye-binding method compared to the fluorodinitrobenzene (FDNB) method. During storage at 50°C, available lysine levels decreased for both methods to about 60% of the original levels (Table 8).

Muchtadi (1986) found available lysine levels to decrease substantially when dendeng was fried. Protease inhibitors were found in stored and fried dendeng but not in boiled meat.

The KCl-soluble N (non-protein N) decreased by 22% during storage of dendeng at 50°C, while it increased slightly for dehydrated meat. Protein solubilised by sodium dodecyl sulphate and B-mercaptoethanol (i.e. denatured protein) decreased by more than 50% in dendeng after 3 months' storage at 50°C, presumably due to stable cross-linking and other reactions probably associated with NEB.

TABLE 8
Available lysine content (g lysine/16 g N) of dendeng and dehydrated meat stored at 50°C (Purnomo, 1986)

Storage time (weeks)	FDNB method		Dye-binding method	
	Dendeng	Dehydrated meat	Dendeng	Dehydrated meat
0	5.7	7.6	4.9	7.2
4	4.4	6.9	3.6	6.1
8	3.6	6.5	2.8	5.9
12	3.3	5.6	2.7	4.8

Texture

Measurements of texture changes (Table 9) parallel sensory increases in toughness or a hardening of texture during extended storage; this is partly due to moisture loss when dendeng is stored in polyethylene bags, but also is related to changes in proteins and possibly NEB reactions.

TABLE 9
Changes in texture of dendeng during storage at 50°C (Purnomo, 1986)

Storage time (weeks)	Deformation force (Newton)	Breaking stress ($N/m^2 \times 10^6$)	Proportion of elongation at break (%)
0	27	0.21	8.6
4	132	1.08	2.6
8	248	2.32	1.2

Organoleptic Quality

Flavour, texture and overall acceptability of dendeng, especially minced fish dendeng, are effected by packaging material and by whether the product has been fried before storage.

Flavour, texture and overall acceptability of fish dendeng decreased during storage at 30°C, and became unacceptable between 9 and 15 weeks at this temperature when stored in plastic bags. Quality was significantly better when product was stored in tinplate cans as representative of an hermetically sealed, light-proof container (Table 10). Whether this is an effect of oxygen, or light, or moisture changes has to be determined, but the effects are noticeable. Vacuum packaging, or the use of oxygen scavengers, would assist but are not practical for these products that are made at the village level.

Frying fish dendeng before storage also significantly affects organoleptic quality (Table 11). Fried fish dendeng is inferior in all attributes to dendeng stored unfried, and fried product stored in polyethylene is significantly inferior to dendeng stored in tinplate cans.

TABLE 10
Effect of packaging on sensory quality of minced fish dendeng stored at 30°C (Purnomo, 1986)

Storage time (weeks)	Flavour		Texture		Overall acceptability	
	Plastic bags	Tinplate cans	Plastic bags	Tinplate cans	Plastic bags	Tinplate cans
0	2.9	2.9	3.3	3.3	3.1	3.0
3	3.1	3.2	3.4	3.5	3.5	3.4
6	3.9	3.0	4.0	3.6	3.8	3.0
9	4.0	3.3	4.1	3.4	4.1	3.3
15	4.6	3.6	4.9	4.0	5.1	4.1

TABLE 11
Effect of frying on sensory quality of minced fish dendeng stored at 30°C (Purnomo, 1986)

Storage time (weeks)	Flavour		Texture		Overall acceptability	
	Plastic bags	Tinplate cans	Plastic bags	Tinplate cans	Plastic bags	Tinplate cans
0	2.9	2.9	3.3	3.3	3.1	3.0
3	3.6	2.6	3.9	3.1	3.9	3.0
6	4.3	2.6	4.4	3.1	4.0	2.8
9	4.3	2.9	4.3	3.2	4.3	3.1
15	5.3	2.8	5.3	3.7	5.5	3.7

Microbiological Quality

Traditional dendeng has a moisture content and a_w too low to permit microbiological spoilage; higher moisture dendeng, however, can spoil due to mould growth. Beef dendeng at 0.89 a_w spoiled in 3 weeks at ambient temperature (15-20°C), and in 6 weeks at 0.85 a_w. Dendeng with an a_w below 0.7 is microbiologically stable, but chemical, physical and organoleptic changes can eventually render the product unacceptable. When stored at 30°C, mould spoilage was evident within 3 weeks at 0.83-0.84 a_w.

Laboratory-produced dendeng (a_w = 0.65-0.7) contained about 10^4-10^5 bacteria/g, about 10^2-10^3 yeasts/g and about 10-10^2 moulds/g. When stored at 25°C, bacterial numbers decreased in 12 months to about 10^3-10^4/g, yeasts could not be recovered after about 8 months and moulds could not be recovered after about 3 months. At 37°C, bacteria decreased in number to about 10-10^2/g in 12 months, yeasts died after 4 months and moulds died after 1-2 months. Thus there is no problem with microbiological spoilage of traditional dendeng during extended storage.

If higher a_w dendeng (a_w = 0.7-0.8) is produced to obtain better palatability, mould spoilage may be possible in the long term. Heating the non-meat ingredients or the spices (10-20 min at 115°C) before dendeng production can decrease substantially the microbial contamination of dendeng before storage, and thus extend the shelf-life. Heating the spices or non-meat ingredients produced no noticeable changes in organoleptic quality (i.e. colour, odour, flavour). Given the much higher microbial loads of the raw materials in developing countries, heating the non-meat ingredients especially the spices, may be a useful method of producing a higher moisture content (and hence higher a_w) dendeng with an acceptable shelf-life. Potassium sorbate (0.1%) prevents mould spoilage of dendeng at 30°C. For developed countries such as Australia, a higher moisture dendeng containing an antimycotic such as sorbate may be an acceptable combination.

SUMMARY

Dendeng is a complex mixture of sliced or minced meat or other high protein food with coconut sugar, salt and spices. For long term storage at tropical temperatures dendeng needs to be packaged to prevent moisture changes; toughening and browning eventually occur to render the product unacceptable. The spice mixture tends to mask rancid flavours in beef dendeng, although fish dendeng does develop rancid-like flavours after 9 weeks at 30°C. Higher a_w dendengs are more palatable but may spoil eventually due to mould growth, but storage life can be extended if the non-meat ingredients are heated before dendeng production or if sorbate is added. The characteristic flavour of dendeng could well appeal to a much wider audience than the South-East Asian consumer.

REFERENCES

Buckle, K.A. & Purnomo, H. (1986). J. Sci. Food Agric. 37, 165.

Muchtadi, D. (1986). Studies on Nutritional Value of Dendeng, an Indonesian Traditional Preserved Meat Product. Food Technology Development Centre, Bogor Agricultural University, Bogor, Indonesia.

Purnomo, H. (1986). Aspects of the Stability of Intermediate Moisture Meat. PhD Thesis. University of New South Wales, Kensington, NSW, Australia.

Purnomo, H., Buckle, K.A. & Edwards, R.A. (1983). J. Food Sci. Technol. (Mysore) **20**, 177.

Purnomo, H., Buckle, K.A. & Edwards, R.A. (1984). J. Food Sci. Technol. (Mysore) **21**, 326.

TRADITIONAL MALAYSIAN LOW AND INTERMEDIATE MOISTURE MEAT PRODUCTS

E.C. CHUAH, Q.L. YEOH AND ABU BAKAR HJ. HUSSIN
Division of Food Technology
Malaysian Agricultural Research & Development Institute
GPO Box No. 12301, 50774 Kuala Lumpur
Malaysia

INTRODUCTION

Malaysia enjoys a rich and varied cuisine, a heritage of the diverse population of the country. There are many traditional products, each originating from a different culture but evolving over the years to gain a unique Malaysian flavour. However, these products can be narrowed down to only a few if only the meat - based products which are in the intermediate and low moisture ranges are considered. The more common traditional intermediate moisture meat products in this country include shredded spiced beef, locally known as sambal daging, spiced meat slice or bak kua in Chinese and Chinese sausage or lup cheong. For low moisture meat products, meat floss is the most well known product.

PRODUCTION TECHNOLOGIES

Shredded Spiced Beef

As the name implies, this product is a highly spiced meat product in the shredded form. The majority of the manufacturers still use the traditional method of production. As the product is in shredded form, coarse fibre meat such as the rump is normally used. The meat is cut into pieces of about 5 x 5 x 10 cm. The pieces are then boiled in water until they are quite tender, drained and allowed to cool. When sufficiently cooled, the fibres of cooked meat are pulled apart manually to obtain strands of meat. Apart from being laborious, this separation involves a great deal of handling. Together with a luke warm temperature, the shredded meat provides a very good medium for microorganisms to proliferate.

While the meat is being shredded, the sauce is prepared. The broth obtained earlier is used in the blending of the ingredients which include coconut milk, sugar, small onion, garlic, ginger, salt, tamarind, dried chilli and coriander. The types and amounts of ingredients used may vary from manufacturer to manufacturer depending on the taste preferences of the consumers.

All the ingredients are poured into an oval-shaped saucepan (kuali) and heat-concentrated with constant stirring to prevent charring. When concentrated to about 50% of its original volume, the shredded meat is poured into the saucepan and stirred continuously until the desired dryness of the product is achieved. Usually towards the later part of cooking, the intensity of the heat is lowered to prevent charring of the product. The manufacturing process could take up to 5 hours, a greater part of this period requiring continuous stirring.

As can be seen from the above method, the most tedious operations are the manual shredding of the cooked meat and the continuous stirring. To overcome these problems, a shredding machine can be used. This has the added advantage of reducing the time required for boiling the meat. In fact it is desirable that the meat to be shredded mechanically be firm for easier shredding. Thus after cooking, the meat is removed from the broth and kept overnight in a chillroom. A certain amount of dehydration occurs and this firms and dries the meat partially.

The shredding machine consists of a bowl chopper with the blunt side of the blades being used for shredding the meat. The desired degree of fineness of the meat fibres will depend on the number of revolutions given. Another device which can be used is a horizontal drum with blunt protrusions from its central rotating shaft. Yet another design uses two spiked drums rotating in opposite directions close to each other. The spikes on the rotating drums are placed in such a manner that they intertwine to give a shredding action to the cooked meat. The meat is fed from the top while the shredded meat falls out from the bottom of the drum.

The tediousness of continuous stirring during the manufacture of the product can be overcome by using a rotary cooker with a scraper and a rotary stirrer. The speed of rotation of both the cooker and the stirrer can be adjusted for better control of the manufacturing process.

The cooker can be heated using liquid petroleum gas (LPG), steam or thermo-oil with thermostatic control. To save cost, some manufacturers also use stationary cookers but with rotary stirrers whose rotation covers the entire cooker. However, the tumbling action of the product is much less, thus requiring a longer drying process. The degree of tumbling not only depends on the rotation of the cooker but also on the design of the rotary stirrer. A schematic diagram of the rotary cooker is shown in Fig. 1. The functions of the scraper and the stirrer are to prevent the product from sticking to the base of the cooker, and to toss and tumble the product to accelerate drying, respectively.

Spiced Beef Slices (Bak Kua)

In Malaysia this product is usually made from pork, chicken or prawn. At the moment, there are no local manufacturers using beef for this product. However, some imported beef products from Taiwan are commercially available.

As the name implies, this product is in sliced form, having a thickness of 2-3 mm. It has a fairly high sugar content and is sometimes known as dried sweet meat or barbequed dried meat.

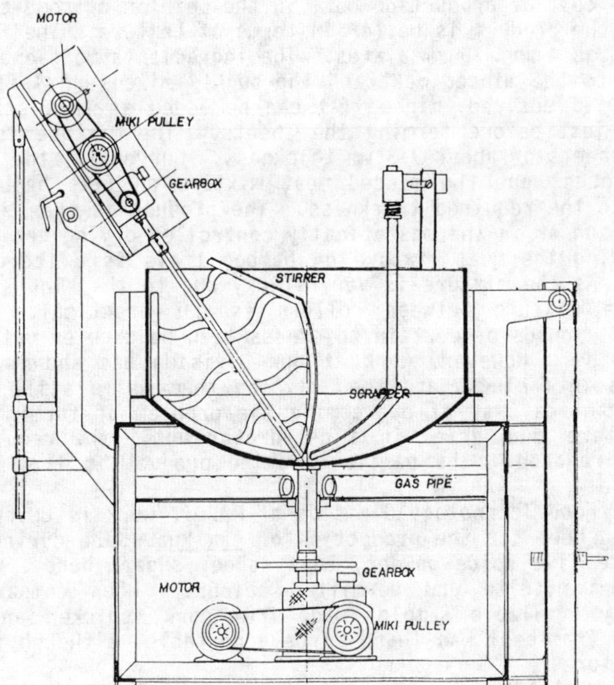

Figure 1. Schematic diagram of a rotary cooker.

Traditionally, this product is made only from pork, usually using ham or shoulder meat. The meat is partially frozen and sliced to the required thickness along the muscle grain using a meat slicer. The sliced meat is then soaked in a curing mixture made up of salt, sugar, soya sauce, monosodium glutamate, spices and permitted colours. After curing for about 2 hours, the slices are placed on slightly oiled bamboo trays and dried at 40 - 70°C for several hours until a moisture content of about 30-40% is reached (Ho & Koh, 1984). After drying, the slices are stacked up and cut into squares. The product is then packed and kept in chillrooms. For longer storage, a freezer is used.

The product is usually sold in ready-to-eat form by grilling the partially dried slices over a charcoal fire until brownish in color. According to Ho & Koh (1984), maltose can be added to give the product a wet and bright appearance. However, the amount of maltose added must not exceed 5%, otherwise the product will become sticky and easily charred during grilling.

Traditional bak kua tends to be a bit tough to bite. The trend nowadays is to make bak kua from minced meat for a more tender product. Using this method, poorer quality meat can be used. This not only reduces the cost of production but, in the opinion of some consumers, the quality of the product is better in terms of texture. The frozen meat is minced through 4 mm - 6 mm plates. The ingredients mentioned earlier are then added to the minced meat and thoroughly mixed and left to cure for 1-2 hours. If desired, minced fat can be added at this stage and mixed thoroughly just before forming the sheets. The mixture is then rolled into thin sheets of about 2-3 mm thickness. Currently, the manufacturers of this product put the minced meat mixture on flat bamboo trays and spread it to the required thickness. The product is then either left to dry in the sun or in thermostatically controlled drying cabinets.

Spreading the meat mixture on bamboo trays is a tedious and messy operation. As the mixture is very sticky due to the high sugar content, passing the mixture between rollers is not practical. Rolling the mixture into moulds of certain thickness can be carried out but again is time-consuming. However, work at our Division has shown that freezing the mixture into blocks and then slicing them while still frozen is an easy and practical way of overcoming the problem of forming the slices. The sheets are then dried in a dehydrator until the required moisture content is reached. The partially dried product is then cut into the desired size.

At the Food Technology Division of MARDI, work is being carried out using minced beef for the production of bak kua. The curing ingredients used include five spice powder, soya sauce, sugar, honey, white pepper, salt, sodium nitrite and permitted colour. When compared with the commercially available samples made from pork, chicken and prawn, the sample made from beef was found to be acceptable, although it had a more compact texture.

A method that has been used in Taiwan for the production of bak kua is by injecting the curing mixture into ham and tumbling it at chilled temperatures for about 6 hours. The temperature of the ham is then brought down to about -12°C and the meat sliced to the required thickness before drying (Chang, 1985). The curing mixture used contains water, sucrose, salt, monosodium glutamate, soya sauce, sodium nitrite, sodium erythorbate, sodium polyphosphate, spices and neucoccine. It is claimed that this method is more time and labour saving besides being more hygienic.

Chinese Sausage (Lup Cheong)

Several grades of Chinese sausages are available commercially. The quality or grade depends on the meat to fat ratio. For liver sausage, the meat is replaced by liver. In this country, pork is normally used for the manufacture of Chinese sausage, although chicken meat can also be used.

The meat used in the manufacture of Chinese sausage is usually ham meat while the fat is from backfat. The meat is manually cut into short strips while the fat is cut into 10 mm cubes. The meat and fat are then mixed thoroughly with the seasoning mixture and are left to marinate for several hours before filling is carried out.

The casing used for Chinese sausage is usually derived from the small intestines of the pig. This type of casing is usually imported in the dry form. These have to be soaked in water before they can be used. With the help of a funnel, the mixture of meat and fat is forced into the casing which is tied at one end. The casing is punctured with needles to allow air trapped during the filling process to escape. After tying and giving a water spray to remove any adhering ingredients, the sausage is allowed to drip for a while before it is dried in a drying cabinet. The heat source is usually burning charcoal. This drying process can continue for up to 3 days, depending on the temperature used.

The traditional method of Chinese sausage manufacture appears to be quite labour-intensive. In some large meat processing factories in this country, certain changes have been made to improve the manufacturing process. Instead of slicing the meat, the latter can be minced through an 8 mm plate. Slabs of backfat which have previously been mixed with salt to remove some water are passed through a strip cutting machine. The fat strips are then passed through the machine again with the strips fed across the series of parallel rotating blades to obtain cubes. However, a dicing machine, which is more expensive, will do a faster and better job using frozen fat. The minced meat and fat are then mixed thoroughly in the seasoning mixture before filling into the casing. The filling process is carried out using an automatic filler with automatic linking facility which can control the weight of each sausage. Pig intestine casing is also replaced by collagen casing which has a more uniform diameter. The amount of heat applied can also be regulated more precisely by the use of modern temperature-controlled drying cabinets. In order to reduce the drying time these cabinets usually incorporate a blower as well as a condenser, the function of the latter being to reduce the humidity in the cabinet. Both these features will reduce the drying period from the usual 72 hours or more to about 35-48 hours.

Depending on the quality of the Chinese sausage required, several formulations are used. Ho & Koh (1984) suggested using a mixture of 63% lean pork and 37% pork fat. This is seasoned with a mixture of ascorbic acid, Chinese wine, light soya sauce, antioxidant, water and curing mixture. The curing mixture used in the above formulation is made up of sodium nitrite (0.5%), sodium nitrate (0.5%) and sodium chloride (99.0%).

Meat Floss

The method of production of meat floss is quite similar to that of shredded spiced beef which has been described earlier. Lean meat is cut along the fibre and cooked in water until it is very soft. The broth is then drained from the meat and kept for later use. The meat is mashed manually into fibrous strips, while the broth is concentrated with spices, sugar and salt added to it. When sufficiently concentrated, the shredded meat is added. The meat mass is then heated at low heat and continuously stirred manually until the desired dryness is achieved. At the final stages of drying, the temperature can be raised to speed up the drying process.

There are commercially available two types of meat floss which differ in texture. Although both have cotton-like appearance, one is more crispy. This is made by adding some vegetable oil to the normal product and a short drying period at a lower temperature is given to achieve the crispy texture. One disadvantage of this product is that there are more 'fines' as the product breaks up more easily when handled due to its more crispy nature. Products made from pork and chicken are commercially available in this country.

In order to increase the production capacity, mechanization has been introduced in several steps of the manufacturing process. Instead of mashing the cooked meat manually, the meat is shredded mechanically. Several designs of the shredder have been used. These shredders are similar to those used in the preparation of sambal daging.

The drying process is carried out in a cooker with a rotating scraper covering the diameter of the cooker. The cooking process is continued until the desired dryness is obtained. The formulation of the above product varies from manufacturer to manufacturer. In general ingredients used are soya sauce, five spice powder, pepper, sugar and monosodium glutamate.

PROBLEMS RELATED TO KEEPING QUALITY

Among the products discussed above, those that fall in the intermediate moisture range are spiced shredded beef, spiced meat slices and Chinese sausage. Their common deteriorative factors are microbiological spoilage (mainly as a result of mould growth) and rancidity due to their relatively high moisture and fat contents.

For shredded spiced beef, samples taken from various markets show that the moisture and fat contents can be as high as 24% and 61%, respectively. With such a high moisture content, coupled with the fact that meat is a good growth medium, microorganisms will proliferate easily unless the product is given some treatment to prolong its shelf-life. This includes lowering its moisture content and/or adding a humectant to lower its water activity (a_w) or heat treatment after packaging to destroy all spoilage microorganisms capable of growing under normal storage conditions. Work carried out earlier by Siaw & Yu (1978) to control mould growth with the addition of 1000 ppm of sorbic acid was not successful. With a moisture content of 21-22% (a_w of 0.7-0.8) the product turned mouldy within 3 weeks of storage at room temperature.

The work by Siaw & Yu (1978) also showed that oxidative rancidity set in relatively fast. Even with an addition of 200 ppm, butylated hydroxytoluene (BHT) was only able to give a shelf-life up to 6 weeks as the rancid taste of the product could be detected by taste panelists after this period. The oil in the product comes not only from the meat but also from the cooking oil and the coconut milk used in the preparation of the product.

Currently, work is on-going in our Division to prolong the shelf-life of this product. The product was prepared using the formulation given earlier. The moisture content was adjusted to about 20% during preparation. An infrared moisture analyser was used to check the moisture content. During preparation, glycerol was added to a level of 4% and 8% based on the weight of the shredded meat. One sample, to which no glycerol was added, served as the control sample. The moistures content of the control, 4% glycerol and 8% glycerol samples after completion of the drying process were found to be 21.2%, 16.8% and 18.5%, respectively, with corresponding a_w values of 0.85, 0.73 and 0.79 while the fat contents were found to be 24.8%, 20.5% and 23.5%, respectively.

The products were packed separately in aluminium laminate pouches and vacuum-sealed. They were then submerged in boiling water until the temperature at the centre of the product reached 95°C. The pouches were then cooled immediately in running water until the internal temperature reached about 37-40°C. After drying, the pouches were stored at room temperature and microbiological, chemical and sensory evaluations were carried out at monthly intervals to determine the condition of the products.

To date, after 6 months' storage at room temperature, all three products were still found to be acceptable with no rancid taste detectable. The moisture content, free fatty acid and peroxide value remained relatively unchanged during the whole storage period. Results obtained are shown in Table 1.

TABLE 1
Changes During Storage of Shredded Spiced Beef

Analysis	Time	Control	+4% glycerol	+8% glycerol
Moisture, %	0 month	21.2	16.8	18.5
	6 months	22.3	17.7	19.6
Free fatty acid, %	0 month	0.8	0.6	0.8
	6 months	0.9	0.7	0.7
Peroxide value, meq/kg	monthly to 6 months	Trace	Trace	Trace

With reference to Table 1, it can be seen that there was no significant rise in the moisture content of the samples upon storage of up to 6 months at room temperature. The free fatty acids (FFA) present are calculated as lauric acid. Since coconut milk was used in the formulation, hydrolytic rancidity will be a significant factor, and the monitoring of FFA is a good indicator of its onset. However, FFA levels remained low throughout the period. Peroxide value is used as an indicator of oxidative rancidity because of the inherent fat present in the meat. From the table it can be seen that only trace amounts are present even after 6 months' storage.

Microbiological examination of the samples showed that Salmonella, Staphylococcus aureus and E. coli were absent in all samples. Yeasts and moulds were not detected in the lowest dilution (10^{-1}) used throughout the 6 months' storage period, while the total viable counts initially were 1.5x10/g, 6.0x10/g and 1.5x10/g for the control, 4% glycerol and 8% glycerol, respectively and remained below 1.0x10/g for the remainder of the storage period. The a_w of the samples also remained fairly constant. The control samples showed a slight increase in a_w from 0.85 to 0.88 after 6 months, while the 4% glycerol sample had an increase from 0.73 to 0.74 and the 8% glycerol sample also showed a negligible increase from 0.75 to 0.76.

This shows that a combination of low a_w of 0.85 and below, mild heat treatment and proper packaging material is effective in prolonging the shelf-life of this product to at least 6 months as compared to 6 weeks even with the addition of antimicrobial agent(s) and antioxidant(s).

Currently, most commercial samples are packed only in low density polyethylene and not vacuum-packed. As such they would be expected to have a shorter shelf-life compared to products with proper packaging and heat treatment. Results of analysis carried out on some commercially available samples are given in Table 2.

TABLE 2
Analysis of some commercial samples of <u>sambal daging</u>

Analysis	Sample				
	K1	K2	K3	M	I
Moisture (%)	5.4	6.0	10.7	16.6	7.6
Fat (%)	42.9	36.0	23.2	15.1	30.7
Protein (%)	26.7	29.5	39.1	40.7	40.4
Ash (%)	6.2	5.9	5.9	8.8	4.1
FFA (%, as lauric acid)	0.2	0.2	0.4	0.4	0.3
PV (meq/kg)	0.1	Trace	0.3	0.3	0.1
a_w	0.48	0.49	0.61	0.56	0.59
Total viable count/g	4.5×10^2	6.5x10	1.2×10^2	6.5x10	1.1×10^2
Yeasts and Moulds/g	3.5x10	0.5x10	<1.0x10	<1.0x10	<1.0x10

Contrary to expectation, the moisture content and a_w of the commercial samples were found to be quite low ranging from 5.4% to 16.6% and 0.48 to 0.61, respectively. However, their oil contents were quite high, ranging from 15.1% to 42.9%, with many of them above 30%. Furthermore, rancid smell could already be detected at the time of purchase in some of the products.

From the analysis data obtained from samples of spiced sliced meat, both made from beef and chicken, the a_w is usually about 0.7-0.9. The product usually has a fat content of about 20-25%. Although these products do not become mouldy as fast as shredded spiced beef, they become rancid quite rapidly unless they are packed in a suitable packaging material. Rancidity is known to set in within 4 weeks of storage at room temperature with normal polyethylene packaging without any visual mould growth. Work is on-going at our Division to determine the shelf-life of this product made from beef and vacuum-packed in aluminium laminates and stored at room temperature.

According to Ho & Koh (1984), Chinese sausages are usually dried to about 50% - 55% of their green weight resulting in the product having a moisture content of about 15-20%. Table 3 below gives some analytical data of the product.

TABLE 3
Chemical composition of Chinese sausage (Ho & Koh, 1984)

	50% drying	55% drying
Moisture (%)	22.9	14.3
Crude fat (%)	42.9	47.3
Crude protein (%)	21.2	23.6
Total ash (%)	4.8	5.3
Sugar (%)	9.3	10.3
Salt (%)	3.3	4.7
Sodium nitrate (ppm)	200-400	200-400
Sodium nitrite (ppm)	< 10	< 10
a_w	Aimed at < 0.61	Aimed at < 0.61

From the above data, it can be seen that the fat content of Chinese sausage is quite high. As such, rancidity will be the main deteriorative factor unless the product is properly packaged or an antioxidant is added. For lower grade products, the fat content is even higher. Although the a_w of the products should be less than 0.6, unless the drying process is carried out sufficiently or a humectant is added in suitable amounts without affecting the taste, the a_w may actually be quite high. Therefore, another possible spoilage factor can arise from the growth of microorganisms, especially moulds.

From discussions held with Chinese sausage manufacturers, products which are sufficiently dry and packed in polypropylene bags will remain acceptable up to about 3 months. Those packed in metal containers will have a longer shelf-life. Therefore to prolong the shelf-life, the product should have an a_w of less than 0.6 to retard the proliferation of microorganisms and be packed in opaque air-tight packaging materials to slow down the onset of rancidity.

Meat floss is usually a relatively dry product and its oil content is also low. With these characteristics its shelf-life is usually much longer than those products in the intermediate moisture range. The moisture and fat contents of the product made from beef were found to be about 16.0% and 3.9% while those from chicken were 13.4% and 9.3%, respectively. The a_w of the beef and chicken floss were found to be 0.6 and 0.62, respectively. A commercial sample of chicken floss which was refried with a bit of oil had an a_w of only 0.28 and a moisture content of 0.4%. With proper packaging, even in clear films with very low water and oxygen permeabilities, the shelf-life of this product should have no problem achieving 6 months or more.

Beef floss prepared at the Division and packed in polyester, nylon and cast polypropylene laminates was stored at room temperature and analysed monthly. The a_w showed a very slight increase from 0.60 initially to 0.62 after 6 months' storage, while the total viable count remained relatively constant at $1.6x10^2$/g at 0 months and $1.0x10^2$/g after 6 months. Yeasts and moulds were not detected at the lowest dilution (10^{-1}) used, while Salmonella, Staphyloccocus aureus and E. coli were also not detected. Sensory evaluation also showed that the product was still acceptable.

CONCLUSION

Traditional Malaysian intermediate moisture meat products such as sambal daging, bak kua and lup cheong as well as low moisture types such as meat floss are still being produced in the traditional way by some small-scale processors. As such, not only is the production capacity low but the shelf-life is also short. Furthermore, the quality of different batches of product may differ due to poor quality control resulting from lack of proper equipment. However, these production problems are slowly being overcome with the introduction of more modern equipment which will not only speed up the rate of production but also result in more consistent quality products due to more precise control of, for example, product particle size, temperature, air speed and other processing parameters. The shelf-life of the intermediate moisture meat products mentioned above varies from about 3 weeks to 3 months, the deteriorative factors being mainly mould growth and rancidity. Their a_w is not low enough to prevent the growth of moulds unless the drying period is prolonged, suitable humectants are added, heat treatment (as a canned product) is given or a combination of two or more of the mentioned treatments is used. Other treatments such as lowering the pH and addition of preservatives can also be considered so long as they do not adversely affect the organoleptic properties of the product or the health of the consumers. Similarly, due to the high fat content of the products, rancidity will set in fairly rapidly unless the products are packed in suitable packaging materials.

Packaging the products under vacuum in appropriate packaging materials will also prolong the shelf-life by retarding microbial growth. At the same time, the absence of oxygen will also reduce oxidative rancidity. With the use of opaque packaging materials, the onset of rancidity is further delayed (Chuah, 1982). Ho & Koh (1984) have also indicated that packaging lup cheong under vacuum may help to increase the shelf-life and lessen the problem of rancidity of the product.

Thus, with the introduction of modern equipment, and the manufacturers having a greater awareness and understanding of the factors that cause deterioration of traditional low and intermediate moisture meat products, greater progress in the manufacture of such products may be anticipated in the future.

REFERENCES

Chang, P.Y. (1985). Improvements in manufacturing of dried pork slices and meatballs in Taiwan. FFTC Technical Bulletin, N90, p. 20.

Chuah, E.C. (1982). Stability studies of sambal daging (unpublished report).

Ho, H.F. & Koh, B.L. (1984). In Proc. 4th SIFST Symposium on Advances in Food Processing, p. 94. 14-15th June, 1984, Singapore.

Siaw, C.L. & Yu, S.Y. (1978). In Proc. MIFT Symposium on Meat Processing Industry in Malaysia, p. 32. 10-12th March, 1978, Kuala Lumpur.

PRESERVATION OF MEAT IN AFRICA BY CONTROL OF THE INTERNAL AQUEOUS ENVIRONMENT IN RELATION TO PRODUCT QUALITY AND STABILITY

ZAK A. OBANU
Department of Food Science and Technology
University of Nigeria
Nsukka, Nigeria

INTRODUCTION

Tropical Africa has a vegetation well-suited to livestock production, and traditional pastoral herdsmen, with age-old experience, in whose hands the bulk of our livestock is kept. According to FAO (1985) livestock production and slaughter estimates (Table 1), Africa has more cattle, goats and sheep than Europe. Comparing absolute numbers of cattle, for example, kept with human population many tropical countries in Africa have over 0.8 cattle/person/annum. Yet the meat industry in Africa remains largely undeveloped and the people's diets shockingly deficient in animal proteins despite huge imports of both raw and processed meats. Mann (1967) and Obanu (1986) have reviewed the problems of meat handling in tropical Africa; some of these problems and their solution hinge on meat processing.

By far the most popular method of meat preservation in tropical Africa is dehydration, especially with solar energy or by smoking. Meat dehydration is traditional in these countries and the products are customarily accepted and desired. The essential feature of dehydration as a food preservation method is that the availability of water, that is water activity (a_w), in the food is lowered to a level at which there is no danger from microbial growth (Scott, 1957) and, in so doing, the water content is reduced to minimize rates of biological, chemical and physical processes which limit the storage life of foods. The quality and stability of dehydrated foods are thus related to the reduction of a_w in these foods. Since the reduction of a_w results from the concentration of the internal aqueous environment, it may be achieved by removing the water as in evaporative dehydration (i.e. drying) and/or by adding solutes to tie up the water as in salting and sugaring.

TABLE 1
Livestock production and slaughter estimates 1974 - 1984 (adapted from FAO, 1985)

	Cattle (in thousands)				Sheep and Goats (in thousands)		
Year	No. Reared	No. Killed	% Slaughter	Year	No. Reared	No. Killed	% Slaughter
IN AFRICA							
1974/76	155,772	18,853	12.10	1974/76	300,616	90,480	30.10
1982	175,357	21,944	12.51	1982	336,372	105,376	31.32
1983	176,229	21,875	12.41	1983	340,584	107,925	31.69
1984	176,206	22,129	12.56	1984	340,958	109,876	32.23
IN EUROPE							
1974/76	134,200	49,205	36.67	1974/76	137,051	78,771	57.48
1982	133,174	46,293	34.76	1982	153,362	85,248	55.59
1983	133,049	46,362	34.85	1983	154,577	85,661	55.42
1984	133,737	47,894	35.81	1984	157,343	87,442	55.57
IN NORTH AND CENTRAL AMERICA							
1974/76	192,684	56,468	29.31	1974/75	36,713	12,130	33.04
1982	197,923	51,002	27.14	1982	36,074	11,725	32.50
1983	194,290	51,973	28.10	1983	35,192	12,190	34.64
1984	193,284	52,704	28.76	1984	34,503	11,836	34.30
IN SOUTH AMERICA							
1974/76	213,690	31,653	14.82	1974/76	121,016	24,852	20.54
1982	243,745	34,349	14.09	1982	125,684	24,285	19.32
1983	245,734	33,004	13.43	1983	124,765	24,413	19.57
1984	248,016	33,036	13.32	1984	127,234	23,189	18.23

TRADITIONAL METHODS, PRACTICES AND PRODUCTS OF MEAT DEHYDRATION

The products of traditional meat dehydration may be of low or intermediate a_w depending on the intensity of water removal or binding which also controls product quality and stability. Product packaging is generally unreliable and practised more for convenient handling than as a preservative measure. Product storage similarly involves no preservative measures such as environmental control or treatment with chemical preservatives. The standard of hygiene at all levels of production, handling and storage is abhorrently poor (Mann, 1967; Obanu, 1986). These obvious defects in our meat processing will be taken for granted, without belittling their significance, in the following discussion of methods, practices and products of meat dehydration in tropical Africa.

Meat Dehydration by Roasting

Roasting of meat over or around an open fire is employed to dry meat and enhance shelf-life. Generally a low-burning or glowing fire is produced by burning hard, rather than soft, woods and the meat pieces singly or in stakes are placed on a wire mesh over or around the fire. More specific details are evident from the processing of tsire (or suya) and balangu - two ready-to-eat popular West African delicatessen meat products that are prepared mostly by the Hausa and Fulani tribesmen.

Tsire or suya - roasted meat pieces : "Suya" is an Hausa word meaning roasted or fire-treated meat. It is in this sense a generic term for partly or fully roasted meat products like kilishi, tsire and balangu. Of these, however, the single product most commonly called suya in the trade is tsire and the two names are used synonymously for the product described hereunder.

Tsire or suya is a widely marketed snack meat product defined as boneless meat pieces staked, smeared with a mixture of salt, spices, groundnut flour and oils, and roasted over or around a low-burning or glowing smokeless fire. Tsire or suya is prepared most commonly from boneless beef and occasionally mutton and goatmeat.

In the production of tsire or suya, meat is boned and cut into chunks about 10 cm long and 8 cm wide. The chunks are later sliced with a curved knife into thin slices about 1 cm thick. The meat slices are staked on to slender wooden sticks about 30 cm long and dusted with a mixture of salt, spices, groundnut flour and groundnut oil for seasoning. The meat stakes are then pinned, or inclined on a wire mesh, around a low-burning/glowing fire at a distance of about 35 cm from the fire. Alternatively, the meat stakes are placed on a wire mesh on top of an oven containing low-burning or glowing fire-wood. Hard wood, low in resins, is used. Roasting takes 20-40 minutes with occasional turning of the meat stakes. The product is displayed unpackaged for sale and wrapped with paper on procurement. It receives no further treatment to enhance stability and its shelf-life is only about 24 h under ambient conditions. As Table 2 (Igene & Abulu, 1984) shows, product moisture is about 20-25%.

TABLE 2
Compositional characteristics of commercial tsire or suya.

Nature of product	Retail net wt of product (kg)	Moisture (wet wt) (%)	Ash (dry wt) (%)	Protein (dry wt) (%)	Total lipid (dry wt) (%)
Raw tsire/suya (before roasting)	0.14+0.02	68.6+1.0	1.2+0.2	21.1+1.1	8.4+1.6
Roasted tsire/suya	0.12+0.01	23.3+2.3	1.6+0.1	59.1+1.4	16.2+2.2

Values are means + S.D for 176 samples each of raw and roasted tsire/suya from 40 processors/vendors (Igene & Abulu, 1984).

Balangu - roasted meat slabs : Balangu may be defined as boneless slabs of lean/organ meat seasoned with salt, spices, groundnut flour and oil and roasted over a low-burning/glowing smokeless fire. Like tsire or suya it is prepared mostly from beef but also occasionally from mutton and goatmeat.

The meat is boned and cut into chunks which are sliced with a curved knife into slabs not less than 1 cm thick - most commonly 4-6 cm or more. The meat slabs are dusted on both surfaces with a seasoning mixture of salt, spices, groundnut flour and oil and then placed on a wire mesh over an oven of low-burning/glowing smokeless fire to roast slowly until done. Roasting takes 30-60 min depending on meat thickness and fire intensity. As for tsire/suya, the product receives no further stabilizing treatment. It is marketed unpackaged and wrapped with paper on procurement. It is similar in moisture content (20-30%) and composition to tsire/suya, and its shelf-life is also about 24 h under ambient conditions.

Smoke-Drying of Meat

Meat preservation by smoking in tropical Africa is by HOT SMOKING which usually involves smoking, drying and high-temperature treatment as opposed to COLD SMOKING which involves dense smoking at low temperature with little or no thermal and drying effects. While in cold smoking distilled (dense, grey, moisture-laiden) smoke from air-limited incomplete combustion is used, hot smoking employs hard wood subjected to complete combustion (with ample air supply) to produce blown smoke which is light and hot containing little moisture; this is, therefore, a cooking-and-drying smoke which causes a lot of shrinkage but produces a cooked smoke-dried shelf-stable product. Meat may be smoked raw or cooked with or without salting. Smoking devices used traditionally vary from smoke pits to one- or multi-chambered ovens with walls made of mud or clay. The methods used also vary in details. The simplest and oldest method of smoking, still used at kitchen-level for meat preservation, is to hang meat slices above an open kitchen fire. Hot-smoking at both domestic and commercial levels is applied to whole carcases of small animals like rodents or to meat pieces.

Smoked whole carcases : Whole carcases of small animals like rodents are singed over fire to burn off the hairs and loosen the horns, claws and hooves for easy removal. The carcases are then eviscerated without skinning. The skin is incised only along the limbs down to the hock, stretched out and held taut with wooden sticks running across the carcase. The carcase is thus stretched into a rhombus to expose a large surface area. For smoke-drying, the carcase is either placed by the fire-side or on a wire mesh over an oven. Smoking could be partial, producing a moist product with 20-50% moisture or full to produce a fully dried product, and drying duration varies from a few hours to 2-5 days depending on the product desired, its size (especially thickness), and the fire intensity. Hot-smoked meat carcases are normally sold unpackaged and their shelf-life varies from 1 week to 1 year depending on the extent of smoke-drying.

Banda - hot-smoked meat pieces : Banda consists of hard smoked pieces of meat mostly from reject cattle, discarded transport beasts like donkeys, horses, asses, camels, buffalo and elephants as well as wildlife. Banda is produced in larger quantities and more widely than any other traditional African dried meat.

The animal, after slaughter, is eviscerated, butchered and the large bones removed. Virtually all of the carcase, including the lean, organs, neckbones, ribs and legs as well as hides/skins and intestines, are used up. These are cut into pieces about 3-6 cm and cooked with a small amount of added water in half-drums for 15-30 min with intermittent stirring. When done the meat pieces are shovelled out and spread on the floor or mat or on wire mesh over the smoking pit or oven to drain-dry. The meat pieces are then smoked with fire generated from burning hard wood in the smoking pit or earthen oven directly below the meat pieces. This involves high-temperature (hot) smoking leading to further cooking, drying and shrinkage. Smoking lasts 18-30 h, depending on meat size and fire intensity, during which the meat pieces are turned periodically to ensure uniform smoke-drying. A mat is used to cover the meat pieces to trap the smoke while the fire intensity is controlled by regulating the quantity of burning wood so as to prevent meat charring. At the end of smoking, the fire is quenched and the meat pieces allowed to cool down to ambient temperature before storage or packaging in sacks and jute/mat bags. The product is dark in colour and of a stone-dry texture with 5-16% moisture and a chemical composition as shown in Table 3. It is a very stable product with a shelf-life of 6-12 months or even up to 2 years under ambient conditons.

Sun-Drying of Meat

Sun-drying of meat is more common in the drier tropics with high sunshine, low humidity and low rainfall for most of the year. In the more humid areas it is restricted mostly to the dry season. Meat for sun-drying is commonly in the form of thin strips or sheets, raw or cooked, and product shelf-life may be entirely dependent on the sun-

TABLE 3
Compositional characteristics of commercial banda

Meat Type	% wet basis			
	Moisture	Protein	Total Fat	Ash
Donkey	9.0	61.3	18.5	8.5
Donkey	5.0	63.0	24.0	8.0
Donkey	4.5	67.7	20.0	7.0
Donkey	10.0	67.4	13.0	7.0
Beef	8.0	68.0	16.0	7.0
Beef	15.4	61.5	18.0	6.0
Beef	16.1	60.4	24.0	8.0
Beef	15.2	63.2	14.5	9.0
Beef	15.5	61.4	14.0	7.0

drying or only partly dependent as in heavily salted or fermented products.

Ndariko - sun-dried meat strips : Ndariko is a Fulfulde (i.e. Fulani) and Hausa name for sun-dried meat with or without a seasoning of salt and spices. It is prepared mostly from beef and occasionally mutton and goatmeat.

The meat is boned and the flesh torn into long strips no more than 2 cm thick. The best products are obtained by tearing the muscles to pieces so that a group of muscle fibres can be dried as a unit. Salt, if applied, is only at seasoning, rather than preservative, level; so also is the use of spices. The meat strips and cleaned intestines, with or without salt seasoning, are hung out in the sun on sticks, ropes and galvanized/barb wire or spread on grass mats to dry. Drying takes usually 6-7 days depending mainly on the weather and, to some extent, the nature and size of meat strips. During drying the meat strips are turned daily, especially if spread on mats, to ensure uniform dehydration. The fully dried product is stored in sacks, pots or metal cans and has a shelf-life of 3-6 months under ambient conditions.

Kilishi - sun-dried roasted meat sheets : Kilishi is a long thin sheet of sun-dried meat seasoned with salt, spices and groundnut flour and lightly roasted. It is a high-priced delicatessen product especially among the Hausa and Fulani tribesmen and the Arabs of Saudi Arabia. It is usually prepared from boneless beef, and rarely from mutton or goatmeat. It is usually made from the hind-leg muscles.

The meat is cut into large lumps that are almost round in shape, and expertly trimmed free of adipose tissue, connective tissue, and veins. The lumps are carefully sliced into continuous thin sheets not more than 2 mm thick and up to 0.5-1.5 m long. The sheets are then spread singly on grass mats placed on the ground or on racks made of sticks in the sun to dry for at least 3.5 h depending on weather conditions. After the first 1-2 h , the meat sheets are turned and pressed on a flat surface (like a table) to smoothen them and further dried for another 1-2 h before the sheets are soaked in a seasoning slurry of salt, spices, flavours and groundnut flour. The sheets are then roasted lightly over a low-burning smokeless fire for 3-5 min and fully sun-dried for another 1-2 h. The product is marketed unpackaged as a high-priced delicacy which is wrapped with paper on procurement. It has a moisture content of 10-15% and a shelf-life of up to 12 months.

Biltong - sun-dried salted meat strips : Biltong, popular in South Africa, is the best documented of all African dried meat products. It is like ndariko in all respects except that it is infused with much more salt by curing.

Boneless meat is torn into strips not more than 1-2 cm thick. The strips are cured by rubbing in salt or by covering them with salt and leaving overnight when a lot of water oozes out of the strips. Spices, such as pepper and ground chillies, are sometimes added; so also is saltpetre, which imparts a bright red (cured-meat) colour to the product. Drying in the sun is carried out by hanging the salted meat strips on strands of galvanized/barb wire. The biltong is ready when a piece, broken or cut off, shows a uniform structure.

Jirge - sun-dried fermented meat strips : Jirge is the Hausa name for the sour sun-dried meat strips prepared and consumed by Shuwa tribesmen. It is a fermented sun-dried meat prepared mostly from beef and occasionally mutton or goatmeat. Jirge is like ndariko in all respects except that it is fermented prior to tearing into strips and drying.

Meat for jirge is boned and cut into chunks which are left until the meat starts to ferment to develop the desired sour taste. It is then torn into strips not more than 2 cm thick and sun-dried with or without addition of salt and spices for seasoning. Drying in the sun is by hanging strips over sticks, ropes and galvanized/barb wire or by spreading over a mat. Drying takes about 5 days, depending on weather, and the meat strips are turned daily to ensure uniform drying. Storage is as for ndariko, in sacks, pots and metal cans, and product shelf-life is up to 6 months or more.

QUALITY AND DETERIORATION OF TRADITIONAL DRIED MEATS

The quality and stability of traditional dried meats, as for other dried foods, depend on the extent of a_w depression and water removal (Labuza, 1971) which vary with the degree of roasting, smoking, sun-drying, salting and fermentation during processing.

Insect Infestation

Insect infestation is a serious problem in all fully dried meats. Initial infestation is usually by dipterian species; maggots and puparia of calliphorid and muscid flies are common on exposed meat during early drying. These are subsequently replaced by dermestid and clerid beetles; the succession of infestation is directly related to changing moisture regimes in the meat. The two principal beetle species most commonly associated with dried meats are Dermestes maculatus and Necrobia rufipes. Beetle infestation occurs during sun-drying and during post-drying cooling of hot-smoked meats when mated females oviposit on the meat. Consequently, the product is usually moved into storage with substantial levels of infestation, especially eggs and early instar larvae. These undergo rapid development under the warm conditions of meat stores. The larvae of both D. maculatus and N. rufipes are voracious feeders, reducing the meat to mere frass within weeks. Control with insecticides should be carefully considered as some are harmful to the consumer. As a routine, all the stores and sacks in which dried meat is kept should be evenly dusted with 0.5% gammexane powder. The use of gammexane, within legal limits, for dusting the meat is recommended, since gammexane evaporates during meat boiling. Application is recommended at 10 ppm on weight basis. If treated meat is not boiled before consumption, the water in which it is soaked should be discarded.

Microbial Stability and Safety

The microbiological stability and safety of traditional dried meats depend, as do other dried foods, on the control of a_w and moisture content below the lower limit at which microorganisms are able to multiply or produce toxins. That the water sorption isotherm shifts with temperature such that at a higher temperature the meat has a higher a_w (Heidelbaugh, 1969) imposes serious microbiological problems when dried meats are produced in colder climates for sale in hotter areas. Also important are consumer-abuse problems under our prevalent insanitary practices (Mann, 1967; Obanu 1986). Of the usual food-borne pathogens only Staphylococcus aureus is able to grow at a_w's as low as 0.86 (Scott, 1957). In addition to its ability to produce one or more potent toxins, Staph. aureus grows rapidly over a relatively wide range of pH; it is frequently found in the nasopharyngeal passages of humans and thus may readily gain access to meat through handlers; it produces toxins of remarkable stability to heat; and it is capable of growth at a_w's which prevent the growth of most other bacteria. However, the minimal a_w for staphylococcal enterotoxin formation in artificial media is somewhat higher than that for growth (Troller & Stinson, 1975). Nevertheless, in the production of dried meat, especially of intermediate a_w, it is desirable that growth of Staph. aureus should be controlled. Even more important than Staph. aureus in dried meat is mould growth. Many dried meats tend to quickly pick up moisture from the atmosphere thereby raising their a_w and moisture contents to levels that encourage mould growth. In the absence of packaging and suitable treatment, mould growth is a serious cause of product deterioration and health hazard from mycotoxins, especially aflatoxin.

Enzymic Deterioration

Enzymes of widespread occurrence such as peroxidases are completely inactivated at a_w's of 0.85 or less (Loncin et al., 1968). However, some hydrolases and lipases are active at a_w's as low as 0.3 or even 0.1 (Acker, 1969). Nevertheless, as currently processed, enzymic deterioration of traditionally dried meats may not be significant.

Non-Enzymic Browning

Among the chemical reactions capable of causing deterioration of dried meats, non-enzymic browning reactions of aldehyde-amino condensation (i.e. Maillard type) is of prime importance. These reactions are strongly water-dependent and reach a maximum at 0.6-0.7 a_w (Loncin et al., 1968). These reactions cause a wide range of defects including darkening, development of scorched off-flavours and odours, toughening in texture, loss of rehydratability, and loss of nutritive value, particularly by lysine destruction. Meat develops a reddish brown discoloration and simultaneously the initial fresh flavour diminishes and is replaced, in the early stages of storage, by a stale flavour followed, in later storage, by unpalatable, bitter, burnt flavours (Sharp, 1957).

Oxidative Deterioration

Lipid oxidation : Lipid oxidation is another major reaction controlling the stability of traditional dried meats. This involves the reaction of oxygen with unsaturated fatty acids producing off-odours. In dehydrated meat an unpleasant "tallowy", "acid", or even "paint-like" odour is developed (Sharp, 1953) and the meat becomes totally unacceptable.

During fat oxidation essential fatty acids are destroyed and free radicals produced which can react with proteins to reduce solubility and biological value (Obanu et al., 1980) as well as destroy fat-soluble vitamins. It is also known that unsaturated carbonyls produced in lipid oxidation can take part in non-enzymic browning reactions (Reynolds, 1965). There is evidence that oxidized lipids, when consumed, may have harmful physiological effects (Schultz et al., 1962).

Protein interactions with oxidized lipids : Formation of peroxides in lipid oxidation leads to reactions with proteins in lipid-protein systems (Schultz et al., 1962; Karel et al., 1975). Carbonyl breakdown products are especially reactive. Among the effects of lipid peroxides on proteins in vitro are loss of solubility due to aggregation or complex formation, chain scission and loss of specific amino acids; the basic and sulphur-containing amino acids are the most susceptible (Obanu et al., 1980).

Organoleptic and Nutritional Quality

Traditional meat dehydration, particularly by hot-smoking, has adverse effects on appearance, odour, texture and taste. Besides processing effects, these attributes of organoleptic quality decline with ambient storage as a result of non-enzymic browning and lipid oxidation as well as lipid-protein and protein-protein interactions induced by intermediate products of lipid oxidation (Obanu _et al_., 1980).

Appearance is an important factor in organoleptic quality and is usually perceived as an integral part of quality. Consumers associate every food with a characteristic colour, and any deviation from this leads to poorer acceptability. Besides, colour change may be directly related to the nutritive value and can be correlated with general quality of dehydrated meat (Obanu & Ledward, 1975).

As regards texture, meat is unique among foods in that its texture is very readily apparent to the average consumer. Even the most uneducated palate can distinguish between tough and tender meat; and between meat which is juicy and meat which is dry. The overall notion of tenderness to the palate includes the impression of wetness produced by the rapid release of fluids during the first few chews. The sustained juiciness is apparently due to slow release of fluids as well as the stimulating effect of fat on salivary flow. Tenderness and juiciness are closely related: the more tender the meat, the more quickly juices are released by chewing and the more juicy the meat appears to be. Dried meat products with reasonable ambient shelf-life are all too dry and hard.

Flavour, embracing taste and odour or aroma, is another important organoleptic attribute. Roasted meat products, i.e. _kilishi_, _tsire_/_suya_ and _balangu_ are very tasty. Besides, lots of spices and flavouring agents are employed to enhance product flavour. Smoke flavour is appreciated among smoked-meat consumers. However, product flavour, unless protected, declines relatively fast during storage as a result of deleterious microbial and chemical reactions.

The nutritional value of dried meat is lowered by harsh or severe processing to enhance shelf-life. These severe treatments set up deteriorative chemical reactions that continue during storage. These reactions, especially non-enzymic browning and lipid-protein interactions are destructive of the component amino acids of meat. Obanu _et al_. (1976) observed that during storage at 38°C of intermediate moisture meat there was a marked decrease in both protein efficiency ratio (PER) and net protein utilization (NPU) from a value similar to that found for fresh meat to a value typical of cereal protein within 24 weeks.

TECHNOLOGICAL PROBLEMS AND PROSPECTS

Industrial Organization

A major problem with traditional dried meat production is that the industry is not organized; processors work in isolation and are ignored even by governments. Often production is shrouded in secrecy and takes place far remote from built-up areas (Okonkwo & Obanu, 1984). There is, therefore, no encouragement from any sector; operational capital and investments are also extremely poor. Production inspection and control are difficult and quality control is ignored. Development of traditional dried meat production necessitates proper organisation of the processors.

Environmental and Personal Hygiene

Abattoirs or, more appropriately, slaughterhouses are located almost exclusively in urban areas; even in these, the scene is busy and dirty, with the air strikingly polluted and heavily charged with both spoilage and pathogenic microorganisms. Refrigeration is almost always non-existent or non-functional, when provided; and meat is left completely exposed to the polluted air, intense sunshine, high ambient temperature and humidity, all of which greatly accelerate the proliferation and metabolic activity of the numerous biological contaminants as well as intrinsic and extrinsic physical, enzymic and chemical changes in meat. Supply of potable water is either totally lacking or grossly irregular or otherwise inadequate. Drains, if provided, are often non-fluent, dirty and stinking. Most often refuse is dumped in heaps close-by; sewage, including human and animal faeces, are similarly treated. Vultures, rodents, flies and several other insects abound as ready vectors that freely infest and infect the meat. Within the slaughterhouse, facilities are minimal, frustrating good abattoir practice. Skinning, evisceration and cutting up of the carcase are often carried out on filthy platforms and tables. Rural areas are completely devoid of abattoirs. At best what exist are sluaghter slabs with roofs over them. In most remote dried meat production units, slaughtering is on bare ground or over leaves/mat. Water is generally lacking and is sparingly used. To talk of personal and meat hygiene is to over-flog an obvious case.

Inferior Raw Materials and Equipment

In all tropical Africa, meat animals are mostly in the hands of traditional pastoral herdsmen found mainly in poorer savannah belts. These herdsmen are to some extent nomadic in search of green pastures. Also they pride themselves on the size of their herd which they guard jealously and hardly part with except when such animals are spent, sick and weak, or when extreme financial need forces sale to meet the need. Under such management, animals are rarely culled for slaughter at the prime stage when they should yield well finished meat with rich flavour at the optimal dressing percentage. Moreover, dried meat processors, having no inducement or control for quality, buy and use the cheapest meat animals available; often the haggard, spent or discarded animals. Thus many will process discarded transport beasts like asses, donkeys and horses and sell these as desired beef products (Okonkwo & Obanu, 1984).

Processing equipment used in traditional meat drying are simplistic and, in some cases, primitive (Okonkwo & Obanu, 1984) making process control difficult, if not impossible. The processor is often at the mercy of the weather for sun-drying or smoke-drying. Product quality is uncertain and variable, so standards are met only by luck and sheer artistry.

Poor Technical Know-how

The process of economic development in any society presupposes a structural, social and technological transformation. For this to be possible in the dried meat industry, new methods, new machines and new/improved products need to be developed to revolutionize aesthetic, qualitative and quantitative production. Unfortunately, however, dried meat production in most parts of tropical Africa is still in its technological infancy in which age-old production methods, materials and machinery remain largely unchanged. A structural transformation of the industry is needed to meet rising demands for better and more hygienically processed meat products. At present, very little is known of the industry, the production technology and needs, product quality, stability and safety, handling, storage and market needs. Process personnel are largely poor and ignorant and need help and education to enhance their know-how, efficiency and productivity for improvement of this industrial sector of our food economy.

CONCLUSION

Although dried meats are traditional to the tropics, the production/marketing sector is largely unorganized in tropical Africa. Several dried meat products have been produced for generations in the region but production has remained either domestic or in isolated units. Little is known of the producers, production methods and the products. Thus, dried meat production in tropical Africa has remained shrouded in secrecy. Under these circumstances production is unregulated and primitive; quality of both raw materials and finished products is variable and inferior; capital investment is frustratingly low; and working conditons are horrible. The situation calls for urgent governmental recognition and action. Nothing short of structural reorganisation with the keen participation of governments, entrepreneurs and researchers can revolutionize and place dried meat production as an industry on a sound footing.

REFERENCES

Acker, L.W. (1969). Food Technol. **23**, 1257.

F.A.O. (1985). Monthly Bulletin of Statistics, Vol. 8, p. 19. F.A.O., Rome.

Heidelbaugh, N.D. (1969). Ph.D. Thesis, M.I.T., Cambridge, Mass.

Igene, J.O. & Abulu, E.O. (1984). J. Food Protect. **47**, 193.

Karel, M., Schaich, K. & Roy, R.B. (1975). J. Agric. Food Chem. **23**, 159.

Labuza, T.P. (1971). CRC Crit. Rev. Food Technol. **2**, 355.

Loncin, M., Bimbenet, J.J. & Lenges, J. (1968). J. Food. Technol. **3**, 131.

Mann, I. (1967). Meat Handling in Under-developed Countries: Slaughter and Preservation. F.A.O., Rome.

Obanu, Z.A. (1986). In Energy, Food and Postharvest Technology in Africa. ANSTI, UNESCO, Nairobi, Kenya.

Obanu, Z.A. & Ledward, D.A. (1975). J. Food Technol. **10**, 675.

Obanu, Z.A., Ledward, D.A. & Lawrie, R.A. (1976). J. Food Technol. **11**, 575.

Obanu, Z.A., Ledward, D.A. & Lawrie, R.A. (1980). Meat Science **4**, 79.

Okonkwo, T.M. & Obanu, Z.A. (1984). In Self-sufficiency in Aninmal Proteins under Changing Economic Fortunes. p. 33. Nigerian Soc. Animal Production, Zaria.

Reynolds, T. (1965). Adv. Food Res. **14**, 167.

Schultz, H.W., Day, E.A. & Sinnhuber, R.O. (eds) (1962). Lipids and Their Oxidation. Avi Publ. Co., Westport, Conn.

Scott, W.J. (1957). Adv. Food Res. **7**, 84.

Sharp, J.G. (1953). Dehydrated Meat. Spec. Report No. 57. H.M.S.O., London.

Sharp, J.G. (1957). J. Sci. Food Agric. **18**, 14.

Troller, J.A. & Stinson, J.V. (1975). J. Food Sci. **40**, 802.

DEVELOPMENT OF INTERMEDIATE MOISTURE TROPICAL FRUIT AND VEGETABLE PRODUCTS - TECHNOLOGICAL PROBLEMS AND PROSPECTS

K.S. JAYARAMAN
Defence Food Research Laboratory
Mysore - 570 011, India

INTRODUCTION

Fresh tropical fruits form an important component of the diets of the people of the tropical regions. Besides being liked for their exotic flavour and colour, they are rich sources of vitamins, especially the provitamin A carotenoids and vitamin C, minerals and carbohydrates and as such constitute an important element of human nutrition.

World production of major tropical fruits of economic importance such as banana, mango, papaya, pineapple and others of lesser importance like guava, sapota, jackfruit, etc. is mostly spread over the tropical and subtropical zones of the less developed countries (Table 1) (FAO, 1983). In the absence of adequate modern facilities for handling, transportation and storage of these highly perishable commodities in these regions there is considerable loss due to spoilage which is aggravated by the high ambient temperatures (22-32°C) and humidity and prevailing unhygienic handling practices.

With specific reference to a fast developing tropical country like India, better farming practices emerging in recent times have resulted in increased agricultural output and also higher production of fruits and vegetables. The production figures pertaining to 1981-82 are given in Table 2 (Bajpai et al., 1985). The present official estimate of production of various types of fruits and vegetables is around 55 million tonnes worth over 60 billion rupees of which the country is presently able to process only less than 1% and an estimated 30%, worth over 20 billion rupees, is lost annually due to lack of adequate low-temperature storage facilities and a proper processing industry. Further, many of the tropical fruits are also prone to chill injury when exposed to low temperatures for long durations. There is, therefore, a growing and urgent need for simple inexpensive processes that would offer a way to save these highly perishable commodities from spoilage under tropical conditions and make them available to regions away from places of abundant production and during lean seasons. This would also fetch a higher price for the grower who would otherwise be compelled to sell the surplus produce at times of glut at unremunerative prices ("distress sale") for want of proper preservation methods.

TABLE 1
World production of some tropical fruits (in 1000 metric tons, 1983 figures)*

	Banana	Mango	Pineapple	Papaya
World	40700	13954	8665	1982
Developing Countries	39885	13933	7714	1921
Africa	4547	836	1257	222
Asia	15708	10917	4768	799
North & Central America	7275	1362	1206	323
Mexico	1624	665	400	230
South America	11529	828	1306	620
Brazil	6692	610	841	460
India	4500	8700	660	270
Phillipines	4200	550	1300	110

* Source : 1983 FAO Production Year book, Vol. 37.

TABLE 2
Production of tropical/semitropical fruits in India during 1981-82 (Bajpai <u>et al</u>., 1985)

Fruit	Area (Lakh ha)	Percentage of total area	Production (Lakh tonnes)	Percentage of total production
Mango	10.02	40.27	86.63	39.08
Banana	3.12	12.54	55.54	25.05
Citrus	2.31	9.28	17.59	7.94
Guava	1.51	6.07	13.15	5.93
Apple	1.54	6.18	9.27	4.18
Pineapple	0.84	3.36	6.43	2.90
Grapes	0.11	0.46	2.22	1.00
Others	5.43	21.84	30.86	13.92

Conventional dehydration and canning processes, which offer the only means of preserving the surplus produce suitable to such regions, have many drawbacks. Specific to fruits and vegetables, the former yields products with rigid structures which need rehydration for prolonged periods and generally have texture and flavour inferior to the fresh materials. Especially for fruits, it is unsuitable due to shrinkage and toughness caused by slow prolonged drying. Canned products, on the other hand, suffer from the disadvantages of bulk, weight, overcooked texture and flavour, high cost (due to high energy input, cost of tinplate and capital investment) and dependence for safety or wholesomeness on the integrity of the container (Brockmann, 1970).

It is now well recognised that if fruits and vegetables are dehydrated to an intermediate moisture level (10-50%), there will be better retention of original flavour and texture compared to fully dried products, with concomitant reduction in bulk, weight and cost of packaging, transportation and storage. Such intermediate moisture foods (IMF) are more appropriate for developing countries in view of their minimal processing requirements, stability under ambient conditions, safety, convenience, ease of nutrient content adjustment, energy savings, and low capital investment (Karel, 1983). They are also eminently more suited than dehydrated or canned foods for military rations (Brockmann, 1970).

In the studies reported in this paper, technological problems encountered in the preparation of IMF with specific reference to tropical fruits and vegetables and methodologies adopted to overcome some of them are discussed. Results of studies conducted on the preparation of shelf-stable IM products from some major tropical fruits and vegetables grown in India using various techniques and their characteristics during storage under different conditions are presented.

TECHNOLOGICAL PROBLEMS IN THE PREPARATION OF IMF FOR TROPICAL REGIONS

Choice of Additives

In all the studies and patents reported so far, the principal method of IMF preparation adopted has been to combine glycerol with other food ingredients like sugar and salt to give products with 15-50% glycerol to achieve a water activity (a_w) less than 0.85 (Karel, 1973). In practice, considerable use has been made of glycerol and propylene glycol which generally have adverse effects on the taste (e.g. sweetness, bitterness, and slight burning sensation) of the food. These and other compounds like sucrose, corn syrup, sorbitol and dextrose have necessarily to be restricted to formulations where sweetness is permissible as for example in fruits, desserts and sweetmeats but they are abnormal and organoleptically unacceptable in the case of products like vegetables.

Similarly, among the common inexpensive food ingredients, salt is the most effective in depressing a_w but, when solely used, the amount needed is excessive and objectionable on flavour grounds. Also, the physiological implications of a high salt intake discourage the use of large quantities of this additive. Further, while it may be combined with other ingredients and limited to the level of normal seasoning in products like vegetables, it is objectionable in fruits.

Although considerable efforts are being made to identify alternative additives, till such time a bland tasting non-toxic a_w-depressing additive is found as a substitute for glycerol, one has to resort to alternative methods of maintaining microbial stability with reduced levels of existing additives to get palatable products.

Economic Considerations

Desorption processing, which is simple and easy to adopt, involving the moist-infusion of fresh fruits/vegetables, normally results in excess residual infusing solution due to abstraction of water from the wet food which, if discarded without reuse, may result in loss of expensive ingredients such as glycerol, thereby making the process economically non-viable and cost-intensive. Recycling of residual infusing solutions requires, besides filtration and decolourisation, their analysis and readjustment for humectant, antimycotic and preservative concentrations by simple rapid methods which can be easily adopted on production lines.

Stability Considerations

It is now well established that reaction rates of lipid oxidation and non-enzymatic browning, the two major reactions influencing palatability and acceptability of foods, are maximum in the IM range (Labuza et al., 1972; Williams, 1976) and both are aggravated by high temperature.

Many of the fruits and vegetables which contain autoxidizable lipids such as the carotenoids (e.g. mango, papaya, and carrot) are sensitive to oxidation and as such face the risk of losing their appealing colour and flavour which decide their organoleptic acceptability and marketability as well as provitamin A value. Similarly, ascorbic acid is also prone to oxidation resulting in loss of vitamin activity besides initiating browning, thereby affecting the colour. Chlorophyll, which is the other major natural pigment influencing acceptability of certain vegetables like green peas, French beans, and capsicum is also prone to rapid loss by conversion to pheophytin in the IM range of a_w. Stabilisation of the unsaturated lipid components as well as ascorbic acid would therefore necessitate the use of antioxidants or oxygen-free packaging.

The plant products are also prone to enzymatic and non-enzymatic browning. Effective control of enzymatic browning calls for blanching which, although suitable for vegetables, seriously affects the texture of fruits at the eatable ripe stage and as such one may have to resort to chemical treatment inhibitory to oxidative enzymes such as polyphenoloxidase. Control of non-enzymatic browning requires low-temperature storage or addition of SO_2 or sulphites. With inadequate low-temperature storage facilities and high ambient temperatures, the use of chemical blocking agents such as sulphites therefore becomes unavoidable although there is increasing consumer interest in wholesome natural foods without additives. Even a substance such as glycerol which is considered indispensable in IMF technology is looked upon with suspicion by the common consumer (Brimelow, 1985).

METHODOLOGIES ADOPTED IN THE DEVELOPMENT OF SOME IM TROPICAL FRUITS AND VEGETABLES AND THEIR RELATIVE MERITS

Methods adopted to overcome some of the technological problems enumerated above, as specifically applied to some common tropical and semitropical fruits and vegetables cultivated in India, are given below.

Fruits

Moist- infusion (desorption) using glycerol and sugar: Aqueous infusing solutions containing predetermined concentrations of both glycerol and sugar calculated to give the desired a_w and preservative concentrations after equilibration were used along with an antimycotic (potassium sorbate) and preservative (SO_2). Although this method minimises processing requirements and cost, and yields products with very good texture and sensory attributes (i.e. soft enough to be consumed as such due to the high concentration of glycerol), the total cost still tends to be abnormally high because of the high cost of food-grade glycerol (thirty times as expensive as cane sugar).

Moist-infusion using sugar only: This is carried out by either two successive soakings in sugar syrups containing the preservatives or by a single soak in sugar syrup followed by partial hot air drying for a short duration to achieve an a_w below 0.85 without significantly altering texture and other attributes. Besides reducing cost considerably, this method would make recycling of residual syrup also much easier since only a single major component, namely sugar, needs to be monitored. This could be done easily and rapidly using a hand sugar refractometer. Preservative concentrations need, however, to be monitored by analysis. Being simple and involving minimal processing and capital investment, this method would be most suited to small-scale industries.

Recycling of spent infusion solutions: Spent solutions containing glycerol and sugar were reused as such or after concentration under vacuum and readjustment of solute and preservative concentration made based on analysis.

Addition of sugar to fruit pulp followed by hot air drying: The solute concentration was raised to a level sufficient to maintain the a_w below 0.7 in an attempt to stabilise the product without the need for an antimycotic. Drying was carried out on materials in the form of thin sheets which could be removed, layered, cut into slabs and packed. While fruits naturally rich in pectin yielded products with soft gel-like structures and moist mouthfeel suitable for direct eating, addition of pectin was resorted to for others deficient in pectin. The method has the specific advantage that expensive ingredients like glycerol and sorbic acid are not needed and many of the accompanying chemical deteriorative reactions likely to be induced by these additives are also avoided. It is simple, economical and least capital intensive and as such is easy to adopt at the small-scale industry level.

Vegetables

Moist-infusion of solutes combined with partial hot air drying: In the case of vegetables, subsequent to soak-equilibration to incorporate the required solutes comprising mainly of salt along with low levels of sugar and glycerol, the product was partially dried to remove a major portion of the moisture to yield an IM product with a moisture content of 50% which could be rehydrated prior to use. The amount of additives required to lower the a_w to 0.85 could thus be brought down to a level that was sufficient to stabilise the product against bacterial spoilage and was tolerable to taste on rehydration. Since vegetables are generally preferred after cooking, such rehydration poses no serious logistical disadvantage. The marked reduction in weight and volume achieved by partial drying would ensure economy in packaging, transportation and storage. By a judicious combination of optimum moisture (to which the product was partially dried so that the texture on rehydration is not adversely affected) and suitable additives (like salt, sugar and glycerol) which are already known to improve rehydration characteristics of dehydrated vegetables, an acceptable IM product could be obtained.

MATERIALS AND METHODS

Raw Materials

The following tropical/semitropical fruits and vegetables procured from the local market were used in the studies :

i) **Fruits**: Banana (Musa cavendishii cv. Dwarf cavendish); mango (Mangifera indica L.cv. Badami); guava (Psidium guajava L. cv. Allahabad safeda); pineapple (Ananas sativus); jamun (Syzygium cumini); jackfruit (Artocarpus heterophyllus L.); papaya (Carica papaya, L.); sapota (Achras zapota L.) and apple (Malus sylvestris cv. Red Delicious).

The fruits were selected at optimum maturity and ripeness suitable for direct eating. Underripe fruits yielded products which were hard and astringent in some cases while overripe ones became pulpy during processing. For preparation of fruit bars, however, fully ripe fruits were used for full flavour development and eating quality.

ii) **Vegetables**: Carrot and cauliflower were specifically chosen since conventional hot air-dried products had poor rehydration characteristics.

Preparation of Raw Materials

Fruits other than pineapple were peeled, cored to remove seeds and cut into transverse or longitudinal slices of approximately 3/8 inch thick. Pineapple, after peeling and coring, was cut into either chunks or rings of 1/2 inch thick. Where fruit was required to be pulped, the peeled and destoned fruit was passed through a pulping and sieving machine using 16-mesh sieve except in the case of jackfruit and guava where the sliced fruit was heated to 90°C with added water before passing through the machine. In the case of fruits prone to enzymatic browning, precautions were taken to prevent the same during preparation by addition of sulphite to water or to the pulps.

Carrot was trimmed, peeled and transversely cut into slices of 1/8 inch thick. Cauliflower was cut into pieces of about 2 inch length and 1 inch thick. They were blanched in boiling water for 5 min.

Additives/Preservatives

Additives/preservatives used in the studies include commercial grade cane sugar and powdered common salt; laboratory grade (BDH) glycerol, potassium metabisulphite (KMS) and dextrose; pectin (apple, 100-150 grade, Marwell, Switzerland); and potassium sorbate (KS) (Koch-Light, England).

Preparation of IM Fruits

Moist-infusion technique using glycerol and sugar: The raw fruit slices were blanched and equilibrated in a solution containing glycerol, 42.25%; sucrose, 42.25%; water, 14.85%; potassium sorbate (KS), 0.45%; and potassium metabisulphite (KMS), 0.2%. The slices were added to the preheated (95°C) soak solution (fruit: solution = 1:2.4), held at 90°C for 3 min with gentle stirring and cooled to room temperature rapidly. The fruit was allowed to equilibrate in the solution at 4°C for 12-16 h, drained thoroughly and packed.

Moist-infusion using sugar only: The raw fruit slices were blanched at 90°C for 3 min in twice the quantity of 70° Brix sugar syrup containing 0.4% KS and 0.2% KMS, cooled rapidly to room temperature, allowed to equilibrate in the solution for 12-16 h at 4°C and thoroughly drained. After filtering and concentrating the residual syrup to 75° Brix under vacuum, the fruit slices were returned to the syrup, resoaked at room temperature for 12-16 h, and then finally drained thoroughly and packed.

Moist-infusion using sugar syrup combined with partial hot air drying : After a single soak in sugar syrup of 70° Brix as described above, the slices having a Brix of about 55° were dried in a cabinet drier at 60-65°C for 1.5 to 2 h to 70° Brix and then packed.

Recycling of spent solutions: Two procedures were tried for reuse of the equilibrated soak solutions :
i) Removal of about 2/3 of the water in the filtered solution by vacuum concentration and readjustment of the solute and preservative concentrations based on analysis; and
ii) Use of a certain proportion of the spent solution to make fresh solution by addition of ingredients based on analysis, the quantity used was limited by its water content.

Direct addition of sugar to pulp and hot air drying: Powdered sugar was added to the fruit pulp to raise the Brix to 30° in the case of mango, papaya, pineapple and jackfruit and to 24° in the case of guava, jamun and banana. After addition of pectin and adjustment of pH where necessary, the pulp was heated to and held at 90°C for 2 min, cooled and treated with 0.2% KMS. It was then spread on aluminium trays (80 cm x 40 cm; 1.0 to 1.5 kg/tray) and dried in a cabinet drier for 1 h at 80°C, 2-3 h at 70°C and 5-6 h at 65°C to a final moisture content less than 15%. The dried fruit gel was removed from the trays as sheets, layered to give a thickness of 12-15 mm, cut into slabs of required size/weight and packed.

Preparation of IM Vegetables

IM vegetables were prepared by moist-infusion of solutes followed by partial hot air drying. The blanched vegetable slices were soaked in an equal weight of a solution containing salt 6.0%, sugar 10.0%, glycerol 5.0%, KS 0.3% and KMS 0.4% for carrot and in twice the weight of a solution containing salt 3%, sugar 10%, glycerol 5%, KS 0.2% and KMS 0.4% for cauliflower for 12-16 h at 4^oC and drained thoroughly. The soaked slices were spread on aluminium trays (1.5 kg/tray of 80 cm x 40 cm), dried in a cabinet drier initially at 55-70^oC to about 55-60% moisture, allowed to equilibrate overnight in a closed container at 4^oC, and then dried again at 60^oC to a final moisture content of 50%.

Analytical Methods.

The proximate composition, reducing and total sugars, acidity, glycerol, sodium chloride, β-carotene and ascorbic acid were determined by the AOAC (1970) methods. Potassium sorbate was estimated by the method of Nury & Bolin (1962) while SO_2 was determined by the iodimetric method (Pearson, 1962). Water activity was determined by measuring the ERH using a modification of the graphical interpolation technique of Landrock and Proctor (Jayaraman et al., 1977).

Storage Studies

IM fruit slices were packed directly in paper (60 gsm)-aluminium foil (0.02 mm) - polythene (LDPE; 75 μ) laminate (PFP) pouches. The fruit bars were first heat-sealed in polypropylene (75 μ) pouches followed by PFP laminate pouches.

IM carrot slices, with and without BHA (applied as a spray dissolved in ethanol at 0.01% level based on the weight of the blanched vegetable after soaking but prior to drying), were packed in SR lacquered cans (301 x 206) under N_2 and in PFP and HDPE (500 G) pouches under air. They were also compressed in cans using gentle pressure by hand with a metallic plunger to approximately twice the original bulk density.

The IM samples packed as above were stored at 0^oC (control), room temperature (RT) (25-30^oC) and 37^oC and periodically examined for colour, flavour and texture (as such for fruits and after rehydration for vegetables) by a panel of judges. They were also analysed by standard procedures for non-enzymatic browning (Hendel et al., 1950), β-carotene and ascorbic acid (AOAC, 1970) and microbiological status (APHA, 1966). Where indicated, browning was also measured as percent diffuse reflectance of the ground sample with a reflectance meter using a magnesium oxide standard to set the instrument to 100% reflectance and in tintometer units using a Lovibond Tintometer.

RESULTS AND DISCUSSION

Characteristics of IM Fruit Slices

The composition and other characteristics of the various IM fruit slices prepared by moist infusion using glycerol and sugar are given Table 3, while the data for those prepared using sugar only are given in Table 4. Organoleptic evaluation showed the products obtained by the different methods to have acceptable colour and flavour and to be soft enough to be eaten as such. Highly acceptable products were obtained from banana, mango, pineapple, jackfruit, apple and guava, while sapota and papaya yielded products lacking in texture and flavour.

In general, IM fruit slices obtained using a combination of glycerol and sugar tended to be superior in texture as compared to those prepared with sugar alone because of the high glycerol content in the former which imparted a moist mouthfeel and soft texture. They also tasted less sweet. Products with sugar alone, especially those obtained by partial hot air drying, were slightly harder, sweeter and stickier. The sweetness due to high sugar-acid ratio observed in some of the fruits processed using sugar alone, especially the low acid ones, could be reduced to the optimum level by incorporation of citric acid in the soak solution at a level of 0.5%.

Reuse of Spent Soak Solutions

Of the two procedures tried for recycling the spent solution, the one based on concentration was found to be ideal since it facilitated reuse of the entire solution. IM fruits prepared by recycling the solution up to two times showed no significant differences in ERH, composition, organoleptic quality and shelf-life as compared to first soak.

Characteristics of IM Fruit Bars

Various factors involved in the preparation of some typical fruit bars and their characteristics are given in Table 5 and their proximate, mineral and vitamin composition in Table 6. ERH studies revealed that the moisture content in the final product should be maintained below 15% to ensure an a_w below 0.70.

In the case of fruits deficient in pectin such as mango, papaya and pineapple, addition of pectin (100-150 grade) at a level of 1.5% on the total sugar content along with adjustment of pH to 4.2 by addition of citric acid was found to improve the texture to a gel-like consistency ideal for direct eating and also help in removal of sheets from trays and in their cutting into slabs. This was, however, not found to be necessary for guava, apple and banana. In the case of jamun, gum arabic at a level of 1% was found to bring about the same effect since adjustment of pH required when pectin had to be used for gel formation adversely affected its taste. Addition of liquid glucose at a level of 6% on total sugar content was found to help in retarding crystallization of sugar in the bars during storage.

Among the fruits studied, mango bars were the best followed by jamun, banana, jackfruit, guava and apple. Combinations of fruits such as mango-pineapple and jamun-banana also yielded acceptable products. Multilayered combinations of different fruits dried individually gave more acceptable products by retaining the individual fruit identities in colour and flavour which were lacking when the pulps of the fruits were mixed before drying.

TABLE 3
Composition and other characteristics of some IM fruit slices prepared by moist-infusion using glycerol and sugar

Parameter	Guava	Pine-apple	Mango	Banana	Jack-fruit	Apple	Sapota
Moisture (%)	36.0	34.8	34.8	30.2	30.0	33.6	38.4
Crude fibre (%)	8.5	1.2	1.8	1.2	1.7	0.8	0.8
Total ash (%)	0.6	0.8	0.4	0.5	0.6	0.8	0.6
Carbohydrates (%):							
Reducing sugars (as dextrose)	3.0	1.9	2.4	4.8	5.2	3.8	9.1
Total sugars (as dextrose)	16.9	25.3	29.6	29.2	29.7	29.3	24.0
Glycerol	37.2	36.2	31.5	36.4	36.8	35.1	35.0
pH	4.4	4.2	4.5	4.9	4.6	5.0	5.0
Acidity (as anhyd. citric acid, %)	0.18	0.45	0.23	0.18	0.15	0.19	0.10
SO_2 (ppm)	224	260	246	301	300	215	360
Pot. sorbate (%)	0.20	0.20	0.20	0.19	0.20	0.21	0.20
β-carotene (mg/100g)	-	-	11.0	-	0.25	-	-
Ascorbic acid (mg/100g)	106.2	6.2	16.6	5.2	2.6	1.0	-
ERH (%)	75.0	79.0	78.0	80.0	70.4	79.5	80.0
Yield (%):							
Raw wt. basis	30	40	35	40	15	50	45
Prepd. wt. basis	50	80	60	70	60	60	65

TABLE 4
Composition and other characteristics of IM fruits prepared by moist-infusion in sugar solution

Parameter	Pineapple	Mango	Banana	Apple	Jackfruit
Moisture (%)	30.2	25.4	23.1	30.7	30.3
Crude fibre (%)	1.2	1.8	1.0	0.8	1.7
Total ash (%)	0.17	0.17	0.60	0.73	0.20
Carbohydrates (as % dextrose):					
Reducing sugars	8.9	8.9	5.5	8.9	9.2
Total sugars	61.0	65.1	62.5	65.0	63.0
pH	4.2	4.8	5.2	5.2	4.6
Brix (degrees)	67	68	65	68	70
Acidity (as % anhyd. citric acid)	0.38	0.13	0.21	0.10	0.15
Sugar/acid ratio	160	501	298	650	420
SO_2 (ppm)	430	369	301	261	300
Pot. sorbate (%)	0.15	0.15	0.16	0.15	0.15
β -Carotene (mg/100g)	-	8.50	-	-	0.25
Ascorbic acid (mg/100g)	6.25	16.60	6.90	1.00	2.6
ERH (%)	78.0	78.5	79.5	78.0	80.0
Yield (%) :					
Raw wt. basis	30	30	40	50	15
Prepd. wt basis	60	50	70	80	60

TABLE 5
Various factors involved in the preparation of some fruit bars and their characteristics

Fruit	Pulp			Bar					
	Yield* (%)	Initial brix (deg.)	Final brix (deg.)	Brix (deg.)	pH	Acidity@ (%)	ERH (%)	Yield* (%)	Texture
Mango	60	17	30	80.0	4.2	1.70	66.6	26	Soft
Banana	67	17	24	80.5	4.7	1.10	67.7	27	Soft
Guava	90	8	24	76.5	3.8	0.85	66.0	40	Soft & Sticky
Papaya	70	17	30	81.5	4.2	0.70	67.0	34	Soft
Jamun	72	12	24	79.0	3.5	0.60	61.4	30	Firm & Sticky
Pineapple	40	16	30	82.5	4.2	1.70	58.1	20	Soft & Sticky
Jackfruit	30	28	30	76.0	4.2	0.80	66.2	10	Soft
Apple	80	9	30	80.5	3.8	1.30	68.4	40	Soft

* On fresh whole fruit weight basis

@ As anhydrous citric acid

TABLE 6
Proximate, vitamin and mineral composition of fruit bars

Fruit bar	Proximate composition (g/100g)						Vitamin & Mineral (mg/100g)				
	Moisture	Fat	Protein	Carbohydrates	Crude fibre	Ash	β-carotene	Ascorbic acid	Ca	P	Fe
Mango	13.4	0.1	2.7	80.4	1.7	1.7	11.20	16.2	58	52	1.4
Banana	9.3	0.1	3.4	81.6	2.4	3.2	-	7.0	47	70	4.5
Guava	12.5	0.1	1.5	77.5	7.0	1.4	-	90.0	60	90	6.0
Papaya	12.1	0.3	1.4	82.6	2.5	1.1	5.6	56.0	100	30	1.1
Jamun	13.1	0.9	2.1	80.7	1.7	1.5	-	-	98	66	2.8
Pineapple	12.0	0.5	1.7	84.6	0.6	0.6	0.43	6.2	57	28	2.5
Jackfruit	13.8	0.3	2.4	77.9	2.8	2.3	0.37	3.1	57	85	1.4
Apple	13.5	0.1	1.3	81.1	2.9	1.1	-	8.0	69	21	2.3

Characteristics of IM Vegetables

ERH, reconstitution and other characteristics of IM carrot and cauliflower slices, prepared using different solute concentrations and combinations combined with partial drying, are given in Table 7. Experiments showed that with salt alone, a level of 6% in the soak solution (giving 9% in the IM product containing 50% moisture) was necessary for carrot and 3% for cauliflower (with a material to solution ratio of 1:1 for the former and 1:2 for the latter) to achieve an a_w less than 0.85. The resultant salt concentration in the IM product reached a tolerable level to taste on subsequent rehydration. However, with salt alone the product was slightly leathery on rehydration.

Salt, sugar and glycerol - the widely used non-toxic a_w depressants - are known to improve the rehydration characteristics of dehydrated vegetables (Neumann et al., 1972; Shipman et al., 1972; Curry et al., 1976). The present studies also showed a definite beneficial effect of sugar and glycerol in improving the texture of IM vegetables prepared by partial hot air drying. While the blanched untreated slices with a moisture content of 88-90% could be dried only to 60% moisture, incorporation of sugar and salt enabled drying to 55% moisture, and a combination of salt, sugar and glycerol permitted drying to 50% moisture without significant shrinkage and impairment of texture.

Sucrose at the 10% level in combination with 5% glycerol in the soak solution gave IM slices with excellent appearance and soft texture which could be rehydrated in 15 min by soaking in boiling water without further heating as compared to 60 min required by low moisture dehydrated slices under similar conditions. Addition of sugar and glycerol along with salt reduced the salt uptake during soaking but the ERH remained below 85% at a moisture level of 50%. The slight sweetness imparted by these additives did not adversely affect palatability after rehydration.

Storage Characteristics

Storage studies showed that with 250 ppm SO_2, IM guava, pineapple, jackfruit and apple slices were acceptable up to 6 months at room temperature (RT) and 3 months at 37°C in PFP. Mango and banana slices remained acceptable up to 9 months at RT and 4 months at 37°C, with a carotene loss of 40% after 9 months at RT and 65% after 4 months at 37°C in mango. Further storage led to browning in the fruits or, in the case of mango, fading of colour. No rancidity or off-flavour was noticed in any of the samples. Browning, measured as OD of the alcoholic/water extract at 420 nm, showed an increase in samples kept at RT and 37°C beyond 4 months except in mango (Table 8). An increase in SO_2 to 500 ppm extended the shelf-life of banana to 12 months at RT and 6 months at 37°C (Table 9).

Compared to IM fruits prepared with a combination of glycerol and sugar, those prepared with sugar alone had generally a shorter shelf-life in all cases except apple. Thus, IM banana and mango were acceptable up to 6 months at RT and 3 months at 37°C with 300 ppm SO_2 and up to 9 months at RT and 4 months at 37°C with 500 ppm SO_2. In the case of apple, the sugar-based product with 500 ppm SO_2 retained acceptability up to 12 months at RT and 6 months at 37°C.

The present studies show that IM fruits made using glycerol and sugar were generally less susceptible to non-enzymatic browning as

TABLE 7

ERH reconstitution and other characteristics of IM carrot and cauliflower slices prepared using different soak solutions and partial drying

Soak solution composition	Characteristics of IM product					
	Moisture (%)	ERH (%)	NaCl (%)	Total sugar (% as dextrose)	Glycerol (%)	Texture on re-hydration
IM Carrot*						
a) 6% Salt	52.0	83.0	11.8	6.3	-	Fair but slightly leathery
b) 6% Salt + 10% sugar	48.0	84.5	10.1	20.0	-	Good
c) 6% Salt + 10% sugar + 5% glyecrol	48.5	83.5	8.9	18.4	7.1	Excellent; As good as freshly cooked
IM Cauliflower**						
a) 3% Salt	43.6	78.0	12.6	8.5	-	Fair but slightly tough
b) 3% Salt + 10% sugar	44.0	82.5	8.8	23.5	-	Good; Softer than (a)
c) 3% Salt + 5% glycerol	55.0	85.0	8.2	-	-	Fair but slightly tough
d) 3% Salt + 10% sugar + 5% glycerol	47.5	80.5	6.9	19.2	9.8	Very good

* Ratio of material to solution = 1:1; Yield on prepared raw material weight basis = 30%; Preservative concentration : KS = 0.15%; SO_2 = 900 ppm.

** Ratio of material to solution = 1:2; Yield on prepared raw material weight basis = 20%; Preservative concentration : KS = 0.15%; SO_2 = 2500 ppm.

TABLE 8
Non-enzymatic browning measured as OD of extract at 420 nm in some IM fruits with 250-300 ppm SO_2 during storage

Temp. (0^oC)	Storage Period (months)	IM Fruit				
		Mango@	Banana	Pineapple	Guava	Apple
0	0-12	0.08(11.0)	0.06	0.05	0.09	0.05
RT (25-30)	2	0.06(8.0)	0.06	0.05	0.09	0.05
	4	0.06(8.0)	0.06	0.07	0.11	0.06
	6	0.06(4.5)	0.05	0.08	0.15	0.08
	9	0.06(4.0)	0.06	0.09*	0.23*	0.08
	12	0.08(3.6)	0.07	-	-	0.09
37	2	0.06(3.5)	0.08	0.05	0.12	0.06
	4	0.06(3.5)	0.10	0.08	0.15	0.06
	6	0.10*(1.0)	0.12*	0.09*	0.31*	0.09*

@ Figures within parentheses represent β-carotene content (mg/100g).

* Product unacceptable due to browning or bleaching in the case of mango.

compared to those prepared with sugar only, at the same level of SO_2. Thus a high concentration of glycerol was found to decrease the rate of browning confirming the earlier findings by Eichner & Karel (1972) and Warmbier et al. (1976). It was also found to significantly reduce oxidation of carotene in mango.

IM fruit bars other than mango prepared with 500 ppm SO_2 retained acceptability up to 6 months at ambient temperature and 3 months at 37^oC. Mango bars were found to retain acceptability up to 12 months at ambient temperature with a loss of about 8% in β-carotene and 23% in ascorbic acid (Table 10) and up to 6 months at 37^oC.

The primary problem in storage of dehydrated carrot is the loss of colour and development of an unpleasant off-flavour (violet-like aroma with bitter soapy after-taste) when stored in the presence of oxygen due to oxidation of β-carotene (Falconer et al., 1964). In IM carrot, these changes were found to occur at a significantly more rapid rate as compared to low moisture carrot. Thus the product packed as such in HDPE was completely bleached within 15 days under ambient conditions while in PFP packs it retained 60% of its original carotene content and its acceptability up to 2 months (Table 11). Compression extended its shelf-life significantly due to exclusion of O_2 from the pack as evident from improved carotene retention and acceptability during storage for 6 months at RT and 4 months at 37^oC. No damage to texture was evident under the conditions of compression and the slices rehydrated normally. Packaging under N_2 gave similar results but was less efficient compared to

TABLE 9
Quality changes in IM Banana prepared using glycerol and sugar during storage in flexible laminate pouches as influenced by SO_2

KMS in soak soln. (%)	Initial SO_2 in product (ppm)	Temp. (^oC)	Storage period (months)	Browning: Abs. at 420nm	Browning: Reflectance (%)	Lovibond Tintometer units Y	R	B	SO_2 (ppm)
0.1	300	0	0-12	0.06	40	6	2.2	1.3	300
		RT	3	0.06	39	6	2.2	1.3	206
			6	0.05	36	6	2.2	1.1	48
			9	0.06	36	6	2.6	1.2	Nil
			12	0.07	34	6	3.0	1.0	Nil*
		37	3	0.08	37	6	2.9	1.3	63
			6	0.12	28	6	3.9	1.4	Nil*
0.2	500	0	0-12	0.06	44	6.5	1.5	-	507
		RT	4	0.08	38	6.5	1.6	-	269
			6	0.08	38	6.5	1.9	0.4	111
			12	0.06	38	6.5	1.6	-	Nil
		37	4	0.08	38	6.5	1.7		79
			6	0.09	30	6.5	3.4	1.9	Nil*
0.3	800	0	0-15	0.05	44	6.5	1.5	-	792
		RT	3	0.05	42	6.5	1.5	-	491
			6	0.05	42	6.5	1.5	-	352
			9	0.05	42	6.5	1.5	-	192
			15	0.09	35	6.5	2.1	-	Nil
		37	3	0.05	40	6.5	1.5		253
			6	0.07	35	6.5	2.1	-	Nil*
			9	0.16	25	6.5	3.9	-	Nil**

* Sample brownish-yellow with weak flavour and of average acceptability.

** Sample brown and caramelised; unacceptable.

TABLE 10
Changes in characteristics of mango fruit bar during storage in paper-aluminium foil-polythene laminate pouches

Temp. (°C)	Storage period (months)	Browning: Extract (Abs. at 420 nm)	Browning: Reflectance (%)	B-Carotene (mg/100g MFB)	Ascorbic acid (mg/100g MFB)	SO_2 (ppm)
0	0-12	0.06	23	12.85	16.25	600
RT	3	0.06	23	12.49	-	600
(25-30)	6	0.06	21	-	-	-
	9	0.06	19	-	-	-
	12	0.07	18	11.83	12.50	525
37	3	0.08	18	12.18	-	-
	6	0.10	11.5	-	-	-

MFB - Moisture free basis.

compression. Coating with BHA had a pronounced effect in retarding oxidative changes and the product remained acceptable in PFP packs under air up to 6 months at RT and 4 months at 37°C. By combining BHA coating with compression, negligible loss of carotene and insignificant off-flavour development occurred during 6 months at RT and 37°C. No browning and adverse changes in texture and rehydration behaviour were evident.

In the case of IM cauliflower, on the other hand, the major problem was observed to be non-enzymatic browning. Samples containing up to 2000 ppm SO_2 were found to remain acceptable only up to 3 months at ambient temperature. A minimum concentration of 3000 ppm SO_2 was found to be necessary to ensure a shelf-life of 6 months.

Microbiological Analysis

The results of microbiological analysis of some of the IM fruit slices prepared using glycerol and sugar, fruit bar (mango) and vegetable (carrot) carried out during storage at three temperatures are given in Table 12. The results show that TPC decreased to negligible levels and the counts for coliforms, staphylococci, yeasts and moulds remained negligible, thus indicating that the products were resistant to bacterial, yeast and mould growth and microbiologically sound for direct consumption. Further studies by Sankaran & Leela (1979) using IM banana and mango on the survival of *Staphylococcus aureus*, *Saccharomyces cerevisiae* and *Aspergillus flavus* revealed that besides reduced a_w, glycerol, KS and KMS independently exerted inhibitory effects on the growth and survival of these microorganisms. The above observations

TABLE 11
Changes in carotene content (mg/100g) of IM carrot slices during storage at different temperatures[a]

Treatment and packing	0°C			RT			37°C		
	2 mth	4 mth	6 mth	2 mth	4 mth	6 mth	2 mth	4 mth	6 mth
A. WITHOUT BHA									
Paper-foil-polythene laminate pouch	14.0 (60.9)	11.5[c] (50.0)	10.0[c] (43.5)	14.0 (60.9)	11.5[c] (50.0)	3.2[c] (13.9)	13.0[c] (56.5)	8.0[c] (34.8)	5.2[c] (22.6)
Compressed in can	23.0 (100.0)	22.5 (97.8)	18.6 (80.9)	22.5 (97.8)	22.6 (98.3)	18.6 (80.9)	18.5 (80.4)	19.4 (84.3)	20.6[b] (89.6)
B. WITH BHA									
Paper-foil-polythene laminate pouch	20.5 (89.1)	20.3 (88.2)	20.3 (88.2)	17.0 (73.9)	16.5 (71.7)	13.5 (58.7)	14.0 (60.9)	14.2 (61.7)	12.2[c] (53.0)
Compressed in can	22.7 (98.7)	22.6 (98.3)	21.8 (94.8)	22.7 (98.7)	23.0 (100.0)	21.8 (94.8)	23.0 (100.0)	20.6 (89.6)	19.0 (82.6)

[a]Initial values : Carotene = 23.0 mg/100g; Potassium sorbate = 0.15% & SO_2 = 900 ppm; Figures within parentheses represent percent carotene retained.

[b]Sample unacceptable due to off-flavour development.

[c]Sample unacceptable due to bleaching and off-flavour development.

TABLE 12

Microbiological data on some IM fruit slices, mango bar and carrot during storage at different temperatures.

IM Product/ Microorganisms	Microbial count (colonies/g)									
		$0^{o}C$			RT($25-30^{o}C$)			$37^{o}C$		
	Initial	2 mth	6 mth	9 mth	2 mth	6 mth	9 mth	2 mth	6 mth	9 mth
MANGO										
TPC	30-80	90	50	15	30-66	25	25	10-40	35	20
Staphylococcus*	30-50	0-30	Nil	Nil	10-30	40	Nil	0-20	26	20
Coliforms	Nil	Nil	Nil	Nil	Nil	Nil	Nil	Nil	Nil	Nil
Yeasts@	Nil	30-40	5	Nil	10-20	10	Nil	Nil	10	Nil
Moulds	Nil	Nil	Nil	Nil	Nil	Nil	Nil	Nil	Nil	Nil
BANANA										
TPC	70	90	105	90	45	75	90	55	40	60
Staphylococcus*	Nil	Nil	5	15	Nil	15	10	Nil	10	10
Coliforms	Nil	Nil	Nil	Nil	Nil	Nil	Nil	Nil	Nil	Nil
Yeasts@	65	Nil	45	Nil	Nil	Nil	Nil	Nil	Nil	Nil
Moulds@	Nil	Nil	Nil	Nil	Nil	Nil	Nil	Nil	5	Nil

TABLE 12 (continued)

IM Product/ Microorganisms	Microbial count (colonies/g)									
	0°C				RT(25-30°C)			37°C		
	Initial	2 mth	6 mth	9 mth	2 mth	6 mth	9 mth	2 mth	6 mth	9 mth
MANGO BAR										
TPC	100-200	100-150	100	-	120-130	100	-	100	100	-
Staphylococcus*	Nil	Nil	Nil	-	Nil	Nil	-	Nil	Nil	-
Coliforms	Nil	Nil	Nil	-	Nil	Nil	-	Nil	Nil	-
Yeasts & Moulds@	5-10	5-10	Nil	-	Nil	Nil	-	Nil	Nil	-
CARROT										
TPC	70	50	50	-	45	15	-	20	25	-
Staphylococcus*	45	15	15	-	10	10	-	15	20	-
Coliforms	Nil	Nil	Nil	-	Nil	Nil	-	Nil	Nil	-
Yeasts & Moulds@	65	30	Nil	-	Nil	Nil	-	Nil	Nil	-

* Staphylococcus wherever noticed were coagulase negative

@ Yeasts - mostly Saccharomyces

Moulds - mostly Rhizopus, A. niger, Cladosporium

confirm the earlier findings by Hollis et al. (1969) who found the plate counts for bacteria, moulds and yeasts in nine IM items to be negligible throughout a storage period of 4 months at 38°C. By inoculation with pathogens, they showed that not only was the growth of these pathogens prevented but their numbers were reduced to negligible levels.

Acceptability Tests

Some of the IM fruits, namely banana, mango and pineapple were used in the form of fruit mixes during several mountaineering expeditions and found to be acceptable and useful. The mango fruit bar was tried as a component of space food during manned space flights and found to be highly relished by the crew.

CONCLUSIONS

There is a growing and urgent need for simple inexpensive processes to preserve fruits and vegetables in developing countries in tropical regions to minimise losses due to spoilage occurring as a consequence of high ambient temperatures and humidity and inadequate facilities for processing, handling, transportation and storage. IMF technology is appropriate for such regions.

Methods based on desorption using infusing solutions containing glycerol and sugar or sugar alone together with antimycotics, or their use coupled with partial drying, could be employed to obtain acceptable shelf-stable IM products from fruits. The spent solutions could be recycled economically after concentration and reinforcement. Alternatively IM fruits could also be made in the form of slabs with soft gel-like texture by addition of sugar and pectin (where necessary) to the pulp and drying. In the case of vegetables, infusion with a combination of salt, sugar and glycerol coupled with drying to 50% moisture could be used to advantage. The products prepared using these techniques were found to have a minimum shelf-life of 6 months at ambient temperature when packed in paper-aluminium foil-polythene laminate pouches. They were resistant to bacterial, yeast and mould growth and microbiologically sound for direct consumption.

ACKNOWLEDGEMENTS

The author is grateful to Dr. T.R. Sharma, Director, Defence Food Research Laboratory, Mysore and Dr. V.S. Arunachalam, Scientific Adviser to the Minister of Defence and Secretary, Department of Defence Research and Development, Ministry of Defence, Government of India, New Delhi for their constant encouragement and kind permission to present this paper. He also gratefully acknowledges the experimental contributions of Mr. D.K. Das Gupta, Mr. M.N. Ramanuja, Mr. B.L. Mohan Kumar and Dr. (Mrs.) Rugmini Sankaran.

REFERENCES

APHA (1966). Compendium of Methods for the Microbiological Examination of Foods (Speck, M.L., ed.). American Public Health Association, Washington, D.C.

AOAC (1970). Official Methods of Analysis, 11th ed. Association of Official Analytical Chemists, Washington, D.C.

Bajpai, P.N., Shukla, H.S. & Chaturvedi, O.P. (1985). In Fruits of India - Tropical and Subtropical (Bose, T.K., ed.), p. 1. Naya Prokash, Calcutta.

Brimelow, C.J.B. (1985). In Properties of Water in Foods in Relation to Quality and Stability (Simatos, D. & Multon, J.L., eds), p. 405. Martinus Nijhoff, Dordrecht.

Brockman, M.C. (1970). Food Technol. **24**, 896.

Curry, J.C., Burns, E.E. & Heidelbaugh, N.D. (1976). J. Food Sci. **41**, 176.

Eichner, K. & Karel, M. (1972). J. Agric. Food Chem. **20**, 218.

Falconer, M.E., Fishwick, M.J., Land, D.G. & Sayer, E.R. (1964). J. Sci. Food Agric. **15**, 897.

FAO (1983). Production Year Book, Vol. 37. Food and Agricultural Organisation of the United Nations, Rome.

Hendel, C.E., Bailey, G.F. & Tailor, D.H. (1950). Food Tchnol. **4**, 344.

Hollis, F., Kaplow, M., Halick, J. & Nordstrom, H. (1969). U.S. Army Natick Laboratories Contract DAAG, 17-67-C-0089.

Jayaraman, K.S., Ramanuja, M.N. & Nath, H. (1977). J. Food Sci. Technol. (India) **14**, 129.

Karel, M. (1973). Crit. Rev. Food Technol. **3**, 329.

Karel, M. (1983). In Chemistry and World Food Supplies: The New Frontiers. CHEMRAWN II (Shemilt, L.W., ed.), p. 465. Pergamon Press, New York.

Labuza, T.P., Cassil, S. & Sinskey, A.J. (1972). J. Food Sci. **37**, 160.

Neumann, H.J. (1972). J. Food Sci. **37**, 437.

Nury, P.S. & Bolin, H.R. (1962). J. Food Sci. **27**, 370.

Pearson, D. (1962). The Chemical Analysis of Foods, 5th ed. J & A Churchill Ltd., London.

Sankaran, R. & Leela, R.K. (1979). J. Food Sci. Technol. (India) **16**, 100.

Shipman, J.W., Rahman, A.R., Segars, R.A., Kapsalis, J.G. & Westcoff, D.E. (1972). J. Food Sci. **37**, 568.

Warmbier, H.C., Schnickles, R.A. & Labuza, T.P. (1976). J. Food Sci. **41**, 528.

Williams, J.C. (1976). In Intermediate Moisture Foods (Davies, R., Birch, G.G. & Parker, K.J., eds), p. 100. Applied Science Publ., London.

USE OF SUPERFICIAL EDIBLE LAYER TO PROTECT INTERMEDIATE MOISTURE FOODS : APPLICATION TO THE PROTECTION OF TROPICAL FRUIT DEHYDRATED BY OSMOSIS

S. GUILBERT
CIRAD (Centre de Cooperation Internationale en Recherche Agronomique pour le Developpement)/CEEMAT (Centre d'Etudes et d'Experimentation du Machinisme Agricole Tropical),
Food Technology Programme, Domaine de Lavalette,
Avenue du Val-de-Montferrand,
34100 Montpellier, France.

INTRODUCTION

The limitations of using reduced water activity (a_w) as the sole preserving factor to formulate intermediate moisture foods (IMF) are numerous (Leistner et al., 1981; Brimelow, 1985). The double necessity of working at microbiologically safe water activities and of keeping a soft texture to assure direct consumption without rehydration implies the use of high concentrations of humectants combined with a relatively high degree of drying.

Combinations of a large variety of humectants, all used reasonably, may be useful since some legal, technological and potential off-flavor problems may be eliminated (Karel, 1976; Guilbert, 1984; Leistner, 1985). But as a consequence, newly developed IMF are often not sufficiently palatable and contain too many additives ("chemical overloading" of the food).

For all these reasons, it may be advantageous to formulate IMF in a water activity range of 0.78-0.86. As for a_w values above 0.78, there is an increasing risk of yeast and mould spoilage and antimycotic agents must be used to enhance preservation. In this case it is desirable to have a surface layer with a high concentration of a given antimycotic agent, therefore permitting a reduction of the overall concentration of additives in the food (Guilbert et al., 1985; Guilbert, 1986). A similar situation is encountered when the addition of an anti-oxygen agent is necessary to protect the food against oxidation from oxygen in the external atmosphere.

Edible protective superficial layers (EPSL) can be used to assure retention and prevent the additive from diffusing freely into the food. Individual protection from external oxygen or from microbial contamination of small pieces or portions of IMF (e.g. IM fruits dehydrated by osmosis) can be achieved, without destruction of the food integrity, by controlling surface preservative concentration with an adequate EPSL enriched with specific additives.

The objective of this contribution is to summarise the state of current knowledge in EPSL technology and its applications through examples, including our recent work, of the types and characteristics of these films, and of the food additive retention and shelf-life extension of IM fruits that can be achieved.

EDIBLE PROTECTIVE SUPERFICIAL LAYERS

The idea of edible protective films and coatings has intrigued packaging and food scientists for a long time. Aside from natural fruit or plant cuticles and waxes, the enrobing of meat products in fatty materials, confectionery coatings, gelatin capsules, sausage casings or wax coatings have also been used for many years. According to Daniels (1973), a package, in the form of a film or a coating, is classified as edible when it is an integral part of the food and consumed as such. It has a dual role as a package and as a food component. Due to this unique property, edible film and coatings have many advantages but also requirements (Table 1). They act as an additional parameter for improving overall food quality and stability. The functional characteristics required depend on the primary mode of deterioration of the food product. EPSL also often reduce and exceptionally suppress, the need for non-edible packaging. Coatings may be applied directly to the surface of the food whereas films are independent structures added to the product.

Since the early work by Schultz et al. (1948) on pectinate films, many materials and technologies have been studied. They have been reviewed comprehensively by Kroger & Igoe (1971), Morgan (1971), Guilbert (1986), Kester & Fennema (1986b), and Biquet (1987). However, most of the information concerning edible food films is found in patent literature. Although 120 British and U.S. patents on the formulation, technology, and application of edible films are given in a book by Daniels (1973), objective and quantitative results are often lacking. Commercial edible films and coatings and their properties and applications have been reviewed in detail by Biquet (1987) and Good (1987).

The majority of edible films and coatings contain at least one component with a high molar mass particularly if a self-supporting film is desired. A wide range of compounds can be used -- amylose, starch-based derivatives, dextrins, carbohydrates, alginates, and other gums, cellulose derivatives, collagen and other protein types, waxes and acetylated glycerides or other fatty materials. They may be used alone or in combination in various forms:simple coating or film, powder, multilayer film, emulsions or microemulsions, etc.

Film formation generally requires a liquid carrier or a solvent, limited to water and/or ethanol for good grade products. The technique of film preparation should be adapted to the characteristics of the film-coating material. Generally, the use of an aqueous solution, colloidal dispersion or emulsion of the film-coating material (and the plasticizer) at a relatively high concentration is often preferred for practical reasons.

TABLE 1
Advantages, Functions and Requirements of Edible Films and Coatings

Advantages	Requirements
* can be consumed directly with the product	* good eating properties (soluble, dispersible in the mouth)
* inexpensive raw material	* organoleptic properties
* can give individual protection to small pieces or portions	* mechanical properties * shelf-life
* can be used inside the food, i.e. between components in a heterogeneous product.	* simple technology * safety
* improve mechanical, organoleptic, nutritional and shelf-life properties of the product by :	* must comply with regulations * good sealings and surface contact
- retarding moisture, gases, fats and oils, or solute migration or transport	* functional under conditions of use
- preventing loss of volatiles (flavors)	
- enhancing food's appearance (glass, sheen surface)	
- preventing packing (non-tacky surface)	
- improving mechanical handling properties	
- contributing to structural integrity	
- protecting from pests and microbial contamination	
- carrying food additives (flavour, colour, antioxidant, preservatives)	

The application and distribution of the film coating material in a liquid form can be achieved : 1) by hand spreading with a paint brush, 2) by spraying, 3) by falling film enrobing, 4) by dipping and subsequent dripping, 5) by distributing in a revolving pan (pan coating), and 6) by bed fluidizing or air-brushing.

Incorporation of divalent or trivalent ions, crosslinking agents or exposure to denaturation conditions by heat or irradiation treatments may be done. The method used to remove the solvent is also important. Excessive heat and/or an excessive rate of evaporation may result in major defects such as formation of cracks, pinholes, slipping, peeling, and non-cohesive films (Banker, 1966; Guilbert, 1986).

Plasticizers such as polyols (sorbitol, glycerol, etc.); mono-, di-, and oligo-saccharides (glucose syrups); and lipids and derivatives (monoglycerides and their esters, phospholipids, etc.) are added to films in order to reduce brittleness and increase flexibility, toughness and tear resistance.

We have been conducting research on EPSL formulation, technology and properties (Guilbert, 1985, 1986; Ricque, 1985; Rojas de Gante, 1985) ; some formulae and characteristics of various films and coatings are given in Tables 2 and 3. Most of those films based on hydrophilic materials cannot be used in the case of foods with a high surface water activity (i.e. $a_w \geqslant 0.94$) because they degrade, dissolve or swell on contact with moisture and then lose their barrier properties. An attempt was made to improve the moisture barrier properties of protein films by using denaturing, cross-linking or tanning agents such as organic acids, tannic acids, tri- or divalent cations, and heat. Denatured protein films are more resistant to moisture but less flexible and transparent (Table 2). The amount and concentration of the denaturing agent used has to be well controlled to avoid an unacceptable acidic or bitter after-taste.

The need for edible films and coatings with good moisture barriers has also led to a study of lipid materials. Foods may be coated with molten wax or fatty materials by direct application, dipping, or pan coating. However, according to our experience, this type of film would raise severe thickness and homogeneity control problems (presence of breaks and pinholes) and their eating properties would be unacceptable (waxy taste and brittle character).

The use of a microemulsion of wax or lipid material has been studied (Guilbert, 1986). The thermodynamic stability of microemulsions makes them uniformly acceptable with time, and the small particle size confers some special performance attribute to the dispersion, such as the ability to deposit a glossy irreversible film after drying. Carnauba wax is known to have unusual performance properties to form oil-in-water microemulsions (Prince, 1960). This kind of microemulsion is extensively used as a fluid, easy-to-use, high-gloss floor polish. A formula for an edible food film based on a very stable oil-in-water emulsion composed of a blend of stearic acid-palmitic acid and carnauba wax and a formula based on a combination of this emulsion and gelatin or casein are given in Table 3. The moisture barrier properties were good but the film formation procedure was delicate to control and it was difficult to obtain a film with constant characteristics.

Recently, Kamper & Fennema (1984a,b; 1985) and Kester & Fennema (1986a) have developed very interesting films of various lipid or waxy materials with a hydroxypropylmethyl cellulose or a mixed methyl cellulose/hydroxypropylmethyl cellulose support. The films have remarkable moisture barrier properties but their application seems

TABLE 2
Formulae and characteristics of protein films

Composition of film-forming liquid[a]		Film characteristics				
First stage	Second stage	Solubility in water: Cold	Solubility in water: Hot	Water barrier properties	Mechanical and organoleptic properties	Observations
Gelatin (250 bloom Rousselot 20%; glycerol 0-10%, water		-	+	Poor	Flexible, smooth, transparent, clear, tasteless, odourless	
	$CaCl_2$ 20%	-	+	Poor	Idem but slight salty and bitter aftertaste	
	Lactic acid 50%	-	+	Fair	Idem but acid aftertaste	
	Tannic acid 20%	-	+	Fair	Brittle, smooth, transparent, brown, astringent aftertaste	
Casein (from Hammarsten, Merck) 10%; NaOH (pH 8); glycerol 5-10%; water		+	+	Poor	Flexible, smooth, transparent, clear, lacteal aftertaste	Brittle without plasticizer
	$CaCl_2$ 20%	+	+	Poor	Idem but slight bitter aftertaste	
	Lactic acid 30%	-	-	Fair	Flexible, slightly rough opalescent, acid aftertaste	Cohesion, when using IR drying or air drying (110 ^{o}C)
	Tannic acid 20%	+	+	Fair	Brittle, smooth, transparent, brown, astringent aftertaste	

TABLE 2 (continued)

Composition of film-forming liquid[a]		Film characteristics				
First stage	Second stage	Solubility in water: Cold	Solubility in water: Hot	Water barrier properties	Mechanical and organoleptic properties	Observations
Casein 5%; gelatin 10% NaOH (pH 8); glycerol 5-10%; water		-	+	Poor	Flexible, smooth, transparent, clear	
	Lactic acid 30%	-	-	Fair	Flexible, smooth, dull, clear, acid aftertaste	Film shrinkage during drying
	Tannic acid 20%	-	+	Fair	Slightly brittle, smooth, brown, astringent, aftertaste	Idem
Serum albumin 10%; glycerol 5%; water		+	+	Poor	Slightly brittle, smooth, transparent, tasteless, odourless	Brittle without plasticizer
	$CaCl_2$ 20%	+	+	Poor	Idem but bitter aftertaste	
	Lactic acid 30%	+	+	Fair	Idem but acid aftertaste	
	Tannic acid	-	-	Fair	Brittle, brown, astringent aftertaste	

TABLE 2 (continued)

Composition of film-forming liquid[a]		Film characteristics				
First stage	Second stage	Solubility in water: Cold	Solubility in water: Hot	Water barrier properties	Mechanical and organoleptic properties	Observations
Ovalbumin 10%; NaOH (pH 8); glycerol 5%; water		+	–	Poor	Flexible, smooth, transparent, clear	Glycerol 5%, optimum, very brittle without plasticizer
	$CaCl_2$ 20%	+	–	Poor	Idem but yellowish	
	Lactic acid 30%	–	–	Fair	Idem	
	Tannic acid 20%	–	–	Fair	Idem but brown	
Soya isolate (Purina) 10%; glycerol 5%; water		–	+	Poor	Flexible, smooth, transparent, clear	
Zein (powder, ICN) 1-2%; ethanol 55-80%; water		–	–	Good	Flexible, grained surface opaque, yellow	
Raw corn proteins (Roquette) ethanol 70-80%; water		–	–		Idem, numerous pinholes	

[a]Composition of the film-forming solutions is given in % (w/w); the setting of the films is achieved by drying (35 ^{o}C, 24h).

TABLE 3
Formulae and characteristics of films

Composition of film-forming liquid		Film characteristics				
First stage	Second stage	Solubility in water: Cold	Solubility in water: Hot	Water barrier properties	Mechanical and organoleptic properties	Observations
Carboxy methylcellulose (CMC) (Hercules) 1-3% ; water		+	-	Fair	Flexible, smooth, transparent, clear, tasteless, odourless	
Malto-Dextrin (DE:3) 3-10% ; water		+	-	Poor	Flexible, smooth, transparent, clear, tasteless, odourless	
Gum arabic (Iranex) 20-30%; glycerol 5-10% ; water		+	+	Poor	Flexible, smooth, transparent, pale brown, tasteless, odourless	Brittle
Sodium alginate (Satialgine, CECA) 2% ; glycerol 20% ; water	$CaCl_2$ 4% water	-		Poor	Flexible, smooth, transparent, clear, tasteless, odourless	Film shrinkage during drying
Sodium alginate (Manucol, Kelco/Ail) 2-5% ; water	$CaCl_2$ 5% water	-		Poor	Idem	Idem
Multi-component films : A (20%) in B (80%) emulsion					Slightly flexible, smooth, opaque, pale yellow, waxy odour and taste	

TABLE 3 (continued)

Composition of film-forming liquid		Film characteristics				
First stage	Second stage	Solubility in water		Water barrier properties	Mechanical and organoleptic properties	Observations
		Cold	Hot			
A : carnauba wax 20%; palmitic and stearic acids 40%; ethanol 40% B : casein 10%; NaOH (pH 8); glycerol 5-7%; water or gelatin 20%; glycerol 5-7%; water		+	+	Good		
A (20%) in B (80%) emulsion	Lactic acid 30%; water	-	-		Idem, acid aftertaste	Film shrinkage during drying
Idem	Lactic acid 30%; tannic acid 20%; water	-	-		Idem, acid and bitter aftertaste	Idem

[a]Composition of the film-forming solution is given in % (w/w); the setting of the films is achieved by air drying (35°C ; 24h).

limited to layered foods (mechanical limitations) and the preparation procedure could be difficult to develop on an industrial scale.

CONTROL OF SURFACE ADDITIVE CONCENTRATION

Using an external layer of high antimycotic or anti-oxygen concentration may allow either the use of the additive without the destruction of food integrity or the use of small amounts of the additive (relative to the total weight) for fabricated foods. Therefore, the diffusivity of the additive within the film and food is of particular importance.

Control of Surface Antimycotic Concentration

The migration of antimycotics, in particular of sorbic acid, is of special interest in a number of cases. Sorbic acid is being used increasingly as one of the numerous "hurdles" employed in the preservation of intermediate moisture foods (Troller & Christian, 1978), low pH foods, or shelf-stable products (Leistner et al., 1981; Leistner, 1985). It is important to be able to predict and control sorbic acid migration between phases during : (a) food treatments (e.g. absorption of sorbic acid during the processing of dried prunes, or loss during cooking of fabricated foods), (b) storage of composite foods (e.g. dairy products or cakes containing pretreated fruits), (c) storage of foods in contact with wrapping materials or films containing sorbic acid (absorption by dairy products covered with paper saturated with sorbic acid), and (d) storage of foods coated with an external edible layer highly concentrated in sorbic acid. In the last two cases, maintaining a local high effective concentration of sorbic acid may allow, to a considerable extent, a reduction of its total amount in the food for the same antifungal effect.

The diffusivity (D) of sorbic acid in various high and intermediate moisture gels (with various substrates and water contents) and in some food products was determined (Guilbert et al., 1985; Giannakopoulos & Guilbert, 1986a,b). This was achieved by tridimensional diffusion in gel cubes or monodimensional diffusion in infinite food columns. For the same substrate concentration by weight, D values of sorbic acid in concentrated sugar solutions decreased slightly when the molecular weight of the sugar was increased. When a liquid subtrate such as glycerol was used, D values (referred to equal concentrations by weight) were higher than in sugar solutions. The diffusion of sorbic acid is related, as a first approximation, to the water content rather than to the water activity of the diffusion medium (Fig. 1).

Diffusivity values appeared to be dependent on the gelling agent (agar-agar) concentration; this indicates that the diffusion should be counteracted by the gel network, probably because the pathlength for diffusion is somewhat increased by the presence of the network. D values of sorbic acid decreased when the temperature decreased (apparent activation energy = 18 kJ mol^{-1}). D values were influenced by the concentration of the diffusant, with a slight decrease when the initial concentration of sorbic acid increased, following a linear relationship.

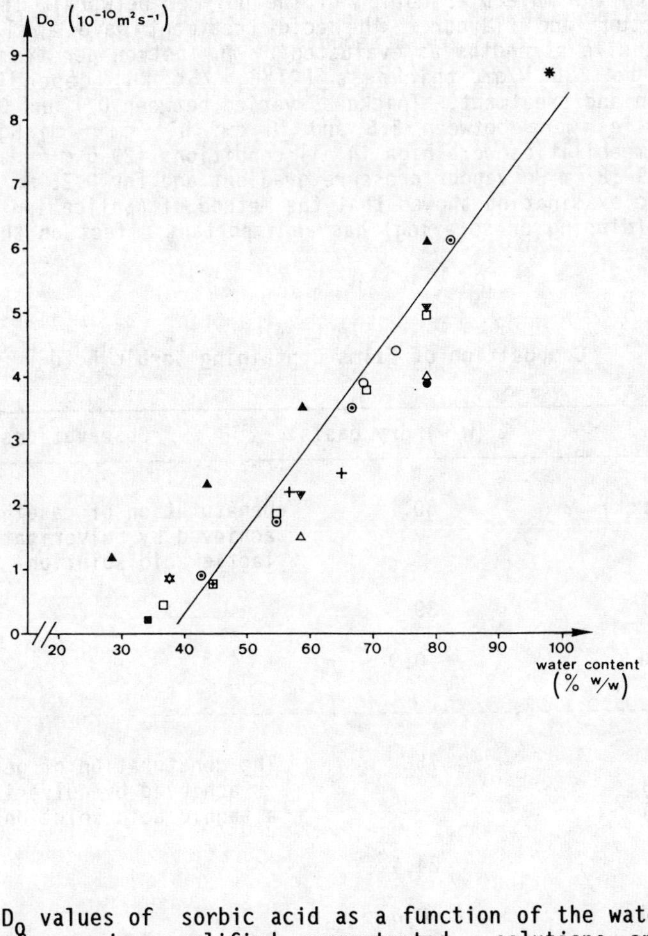

Figure 1. D_0 values of sorbic acid as a function of the water content of various gelified concentrated solutions and of red currant jelly, processed cheese and chestnut cream. Regression line given is calculated from the gelified concentrated saccharose solutions.

✳ water	▽ P.E.G. 600
□ sucrose	○ sucrose + gelatine
⊙ glucose syrup DE 46	+ sucrose + peanut oil
△ maltodextrin DE 20	⊞ processed cheese
● maltodextrin DE 6	✿ redcurrant jelly
▲ glycerol	■ chestnut cream

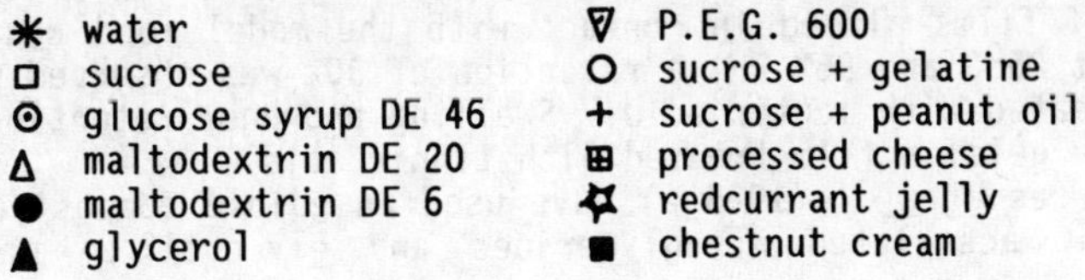

The retention of sorbic acid in gelatin and casein films treated with, respectively, tannic or lactic acid, placed over a model food system (saccharose 44%, agar 1.5%, water 54.5%, 0.95 a_w) was also investigated. The formulae of the films used are given in Table 4. Denaturing or crosslinking treatments were achieved by pulverisation of tannic or lactic acid during a second stage in order to have a better retention of the antimycotic agent and to limit the swelling of the film (by reducing the molecular motion of the polymer network). The films had good structure and flavour. The acid treatment gave a slight after-taste. Tensile strengths as evaluated by an Instron penetrometer ranged from 0.2 to 2.5 N/mm thickness (21°C, 75% RH) depending on the formulation and treatment. Thickness varied between 0.1 and 0.4 mm. Air permeabilities were between 2.5 and 18 $cm^3\ h^{-1}\ cm^{-2}\ cm\ Hg^{-1}$. Water vapour permeabilities were high in all conditions (29 $g\ m^{-2}\ day^{-1}\ mm\ Hg^{-1}$ at 30°C, 29-18 mm Hg vapour pressure gradient and for 0.25 mm thickness). Microscopic examination showed that the method of application of the acid treatment (dipping or spraying) has an important effect on the structure of the film.

TABLE 4
Composition of Films Containing Sorbic Acid

Component	% (w/w) dry basis	Observations
Casein (Merck)	60.1	Denaturation of casein films is achieved by pulverisation of a lactic acid solution, 30% (w/w)
Glycerol	39	
Sorbic acid	0.9	
Gelatin (250 Bloom) (Rousselot)	75	The denaturation of gelatin film is achieved by pulverisation of a tannic acid solution, 30% (w/w)
Glycerol	24.4	
Sorbic acid	0.6	

Fig. 2 shows sorbic acid retention as a function of time for the two types of films placed in contact with the model food system. After 35 days (at 25°C and 95% RH) a retention of 30% was observed with the casein film treated with lactic acid. Swelling and poor retention were observed with the gelatin film treated with tannic acid.

Torres _et al_. (1985 a) have used an edible composite film composed of zein, acetylated monoglycerides and glycerol to maintain a high concentration of sorbic acid at the surface of an IM cheese analogue. Sorbic acid distribution experiments have shown that the diffusion of the preservative into the food bulk was significantly retarded by the coating. Sorbic acid diffusivity was estimated between 3 and 7 x 10^{-9} cm^2 s^{-1}, i.e. 150-300 times smaller than in the food bulk.

Control of Surface Anti-oxygen Concentration

We have measured the retention of α-tocopherol in gelatin films [gelatin, 250 bloom, 76% (w/w) dry basis; glycerol, 19% (w/w) dry basis; and DL-α-tocopherol, 5% (w/w) dry basis] treated or not by pulverisation with a tannic acid solution (30%, w/w). The film's properties were described earlier in this review. The films were eventually placed in contact with an aqueous model food system (saccharose 44%, agar 1.5%, water 54.5%) or with margarine. Due to the low solubility of tocopherol in water, its diffusivity value in an aqueous model food was extremely low. In a fatty material the apparent diffusivity of tocopherol (as evaluated by monodimensional diffusion in infinite columns) ranged from 10 to 30 x 10^{-7} cm^2 s^{-1}.

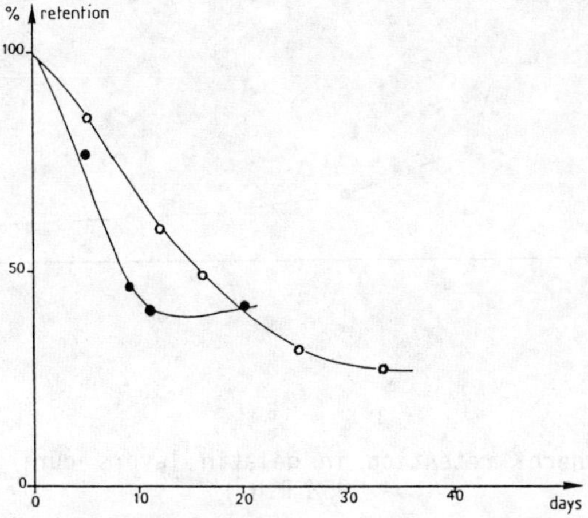

Figure 2. Sorbic acid retention in a protein layer in contact with a model gel (a_w = 0.95)

● gelatin layer treated with tannic acid

○ casein layer treated with lactic acid

Retention rates during storage (25°C, 95% RH) are given in Fig. 3. We observed 1) an evolution of measured tocopherol similar for isolated layers and layers in contact with the model food, 2) a rapid decrease of tocopherol content in the gelatin layer in contact with margarine, but 3) almost no relative loss of tocopherol when the gelatin layer placed in contact with margarine had been previously treated with tannic acid. This treated gelatin layer enriched with tocopherol could be used to reduce the oxidative rancidity and deterioration in some food products but more experiments are necessary in order to have a better understanding and control of the system.

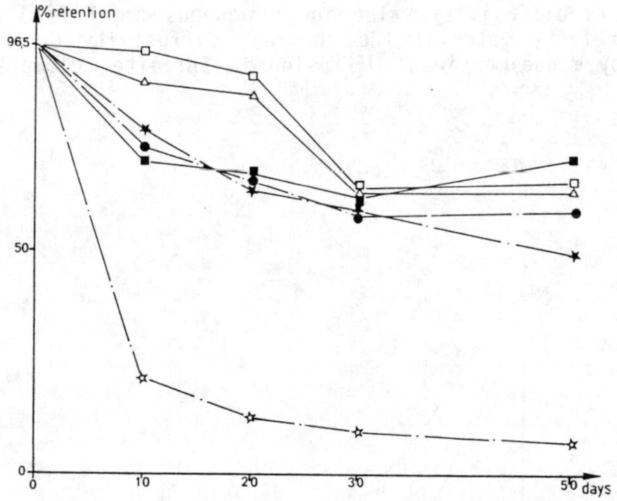

Figure 3. Tocopherol retention in gelatin layers during storage (25°C, 95% R.H.)

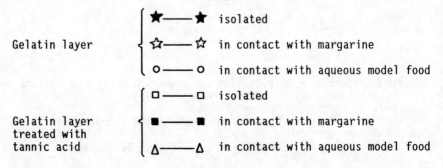

Gelatin layer	★——★	isolated
	☆——☆	in contact with margarine
	○——○	in contact with aqueous model food
Gelatin layer treated with tannic acid	□——□	isolated
	■——■	in contact with margarine
	△——△	in contact with aqueous model food

INDIVIDUAL PROTECTION OF INTERMEDIATE MOISTURE FRUIT

The market for IM fruits is presently growing. They can be consumed directly as "fruit snacks" or in combination with nuts, breakfast cereals, etc. They can also be incorporated in various fabricated foods such as biscuits, cookies, and ice creams. Most of these IM fruits (papaya, mango, pineapple, jackfruit, etc.) are produced in South Asia. Schematic processing is given in Fig. 4. In the usual process, products are dried after soaking in a sugar solution; the final products (15% moisture, 0.6-0.7 a_w) have a hard chewy texture. In some cases, it would be desirable to obtain softer products with a higher water content which would look "fresher or more natural".

Using a moderate drying process (Fig. 4 - proposed process) should lead to lower costs and better quality fruits. However, at a_w values above 0.78 there is an important risk of superficial yeast and mould spoilage and of oxidation of vitamins, flavour compounds and pigments. Therefore, adequate additives must be used to stabilize the product and extend its shelf-life.

IM papaya and apricot cubes (with different water activities ranging from 0.70 to 0.90), were coated with a casein or carnauba wax EPSL containing sorbic acid in order to prevent the growth of surface microorganisms.

IM papaya and apricot cubes (1 x 1 x 1 cm) from commercial origins (some of their characteristics are given in Table 5) were first equilibrated at 30°C for 6 days with different saturated salt solutions (respectively $SrCl_2$ - NaCl - KCl - $BaCl_2$) at 70.8, 75.3, 84.3, 90.3% relative humidity. Once equilibrium was reached, the cubes were coated by dipping in a casein solution (at pH 8 for solubility reasons) with glycerol (as a plasticizer) or in molten carnauba wax with or without sorbic acid.

Coated IM fruit cubes and non-coated samples (control) were stored for 1 day at their equilibrium relative humidity before being inoculated by immersion with osmoresistant microorganisms (*Aspergillus niger* or *Saccharomyces rouxii*). The samples, in triplicate, were incubated at 30°C and at their equilibrium relative humidity for 40 days.

Sample composition and time for apparent development on at least one of the three replicates of *A. niger* or *S. rouxii* are given in Tables 6 and 7. Limiting a_w levels for growth of *S. rouxii* and *A. niger* in laboratory media are 0.62 and 0.77, respectively (Richard-Molard *et al*., 1986; Richard-Molard & Lesage, 1986). *S. rouxii* is often found in products with high concentrations of sugar such as jams, syrups or dried fruits. However, no spoilage was observed for our inoculated samples with a water activity below 0.75 after 40 days of incubation (Tables 6 and 7). At higher a_w, the delay for a first apparent spoilage depended on the type of coating used. For example, the casein coating without sorbic acid had no significant effect on the stability of either the papaya or apricot cubes with 0.84 or 0.90 a_w, i.e. rapid spoilage was observed. In the case of the wax coating without sorbic acid, no apparent growth occurred if no cracks or holes were present in the coating.

Significant improvement of the microbial stability was observed with coatings containing sorbic acid and no spoilage could be detected after 40 days' storage in the case of papaya and apricot cubes coated with carnauba wax containing sorbic acid. The coatings efficiencies were in

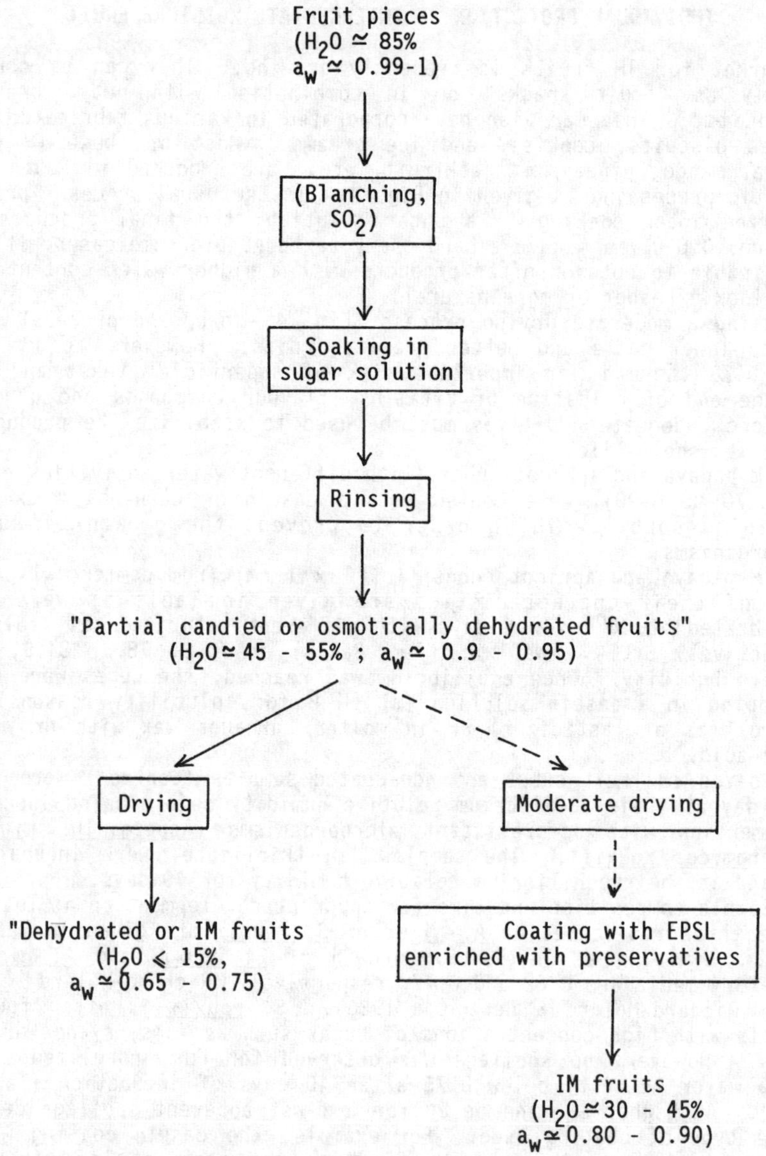

Figure 4. Schematic processing for IM "partial candied" fruits pieces or cubes.

⟶ usual process

- - -> proposed process

TABLE 5
Characteristics of IM papaya and IM apricot cubes

Characteristics	Papaya	Apricot
pH (dilution procedure)	3.95-4.05	4.01-4.21
SO_2 content (ppm, dry basis)	500-800	450-800
Sugar content (HPLC)(% dry basis):		
sucrose	36.0	10.4
glucose	18.1	35.0
fructose	15.9	10.4

the following order : carnauba wax + sorbic acid > carnauba wax > casein + sorbic acid > casein ⩾ no coating. The fact that casein + sorbic acid EPSL was less effective than the carnauba wax + sorbic acid EPSL may be explained by the poor retention properties and the high initial pH of the casein film. Further treatment of the coated surface with an acid (e.g. citric or lactic acid solution applied by spraying or dipping) can be done in order 1) to increase the rigidity of the gel network and 2) to reduce the pH and to increase the surface availability of the most active form of sorbic acid (undissociated acid). But the surface pH reduction obtained is temporary and some modifications of the surface appearance can be noticed (less sheen and transparent films) depending on the water content of the coating when the acid application is made.

The possibility of obtaining a permanent reduction of surface pH was explored by Torres et al. (1985b). Enhanced stability against surface growth of *Staphylococcus aureus* in an IM cheese analog was obtained by enrobing it in a carrageenan-agarose gel matrix containing sorbic acid (Torres & Karel, 1985). The effect on the surface pH was described using a Donnan equilibrium model that predicted a permanent pH difference between the food bulk and the surface containing immobilized negatively charged macromolecules. Torres & Karel (1985) also improved the stability of the same IM cheese analogue by using the mixed zeinacetoglyceride coating previously mentioned (Torres et al. 1985 a) containing a high concentration of sorbic acid. Surface challenge experiments with *S. aureus* confirmed the increased resistance to surface microbial growth.

CONCLUSION

The application of an external layer with a high antimycotic or anti-oxygen concentration may allow either the use of the additive without the destruction of food integrity or the use of small amounts of the additive (relative to the total weight) for fabricated foods. Therefore, the diffusivity of the additive within the film and food is of particular

TABLE 6

Stability of coated IM papaya cubes (with a_w ranging from 0.70 to 0.90) inoculated with *A. niger* or *S. rouxii*

Sample	Composition % (w/w) dry basis					HRE % of the sample during storage	Delay for first apparent spoilage (days)	
	Fruit	Casein	Glycerol	Wax	Sorbic acid		A. niger	S. rouxii
Control (non coated)	100					70.8	> 40	> 40
						75.3	> 40	> 40
						84.3	13	3
						90.3	4	1
Coated with casein EPSL	98.75	0.75	0.5			70.8	> 40	> 40
						75.3	> 40	> 40
						84.3	12	2
						90.3	4	2
Coated with casein + sorbic acid EPSL	98.7	0.8	0.5		8×10^{-3}	70.85	> 40	> 40
						75.3	> 40	> 40
						84.3	22	> 40
						90.3	10	17
Coated with carnauba wax EPSL	99.93			0.07		70.8	> 40	> 40
						75.3	> 40	> 40
						84.3	> 28	> 40
						90.3	10	14
Coated with carnauba wax + sorbic acid EPSL	99.89			0.11	3.5×10^{-3}	70.8	> 40	> 40
						75.3	> 40	> 40
						84.3	32	> 40
						90.3	10	> 40

TABLE 7
Stability of coated apricot cubes (with a_w ranging from 0.70 to 0.90) inoculated with *A. niger* or *S. rouxii*

Sample	Composition % (w/w) dry basis					HRE % of the sample during storage	Delay for first apparent spoilage (days)	
	fruit	casein	glycerol	wax	sorbic acid		*A. niger*	*S. rouxii*
Control (non coated)	100					70.8	> 40	> 40
						75.3	> 40	> 40
						84.3	17	5
						90.3	13	1
Coated with casein EPSL	98.7	0.8	0.5			70.8	> 40	> 40
						75.3	> 40	> 40
						84.3	12	6
						90.3	4	2
Coated with Casein + sorbic acid EPSL	98.4	1.0	0.6		1×10^{-2}	70.8	> 40	> 40
						75.3	> 40	> 40
						84.3	19	> 40
						90.3	17	> 40
Coated with carnauba wax EPSL	99.9			0.1		70.8	> 40	> 40
						75.3	> 40	> 40
						84.3	19	> 40
						90.3	6	21
Coated with carnauba wax + sorbic acid EPSL	99.85			0.15	5×10^{-3}	70.8	> 40	> 40
						75.3	> 40	> 40
						84.3	> 40	> 40
						90.3	8	23

importance. Edible protective superficial layers (EPSL) can be used to assure the retention of the additive and to prevent it from diffusing freely into the food.

Various casein- and wax-based EPSL enriched with sorbic acid as an antimycotic agent were tested to improve the surface microbial stability of intermediate moisture fruit cubes (apricots or papaya) inoculated with Aspergillus niger or Saccharomyces rouxii. It appears that significant improvement of the shelf-life of IM fruits with an a_w of 0.84 can be obtained when EPSL containing sorbic acid are used. Therefore, since the depression of water activity to a level below 0.85 without excessively impairing organoleptic properties is often very difficult, the use of antimycotics retained in EPSL can be an effective and safe manner to impart protection to IM fruits against microbial growth. The possibility of producing stable, low cost IM fruits (with $a_w \simeq 0.85$) by osmotic dehydration and subsequent application of EPSL will be explored in detail in future studies.

REFERENCES

Banker, G.S. (1966). J. Pharm. Sci. **55**, 81.

Biquet, B. (1987). Moisture Transfer in Foods and Edible Moisture Barriers. M.Sc. Thesis, Minnesota Univ., St Paul, Minn.

Brimelow, C.J.B. (1985). In Properties of Water in Foods in Relation to Quality and Stability (Simatos, D. & Multon, J.L., eds), p. 405. Martinus Nijhoff, Dordrecht.

Daniels, R. (1973). Noyes Data Corporation, Park Ridge, N.J.

Giannakopoulos, A. & Guilbert, S. (1986a). J. Food Technol. **21**, 339.

Giannakopoulos, A. & Guilbert, S. (1986b). J. Food Technol. **21**, 477.

Good, D. (1987). Edible films in the film industry. Frontiers (Diversified Research Laboratories).

Guilbert, S. (1986). In Food Packaging and Preservation (Mathlouthi, M., ed.), p. 371. Elsevier Applied Science Publishers, London.

Guilbert, S. (1984). In Additifs et Auxiliaires de Fabrication dans les Industries Agro-Alimentaires (Multon, J.L., ed.), p. 199. TEC et DOC Lavoisier, Paris.

Guilbert, S. (1985). Effets de la Composition et de la Structure des Aliments sur l'Activite et la Mobilite de l'Eau ou de solutes. These de Doctorat d'Etat. Universite des Sciences, Montpellier, France.

Guilbert, S., Giannakopoulos, A. & Cheftel, J.C. (1985). In Properties of Water in Foods in Relation to Quality and Stability. (Simatos, D.7 & Multon, J.L., eds), p. 343. Martinus Nijhoff Publ., Dordrecht.

Kamper, S.L. & Fennema, O.R. (1984 a). J. Food Sci. **49**, 1478.

Kamper, S.L. & Fennema, O.R. (1984 b). J. Food Sci. **49**, 1482.

Kamper, S.L. & Fennema, O.R. (1985). J. Food Sci. **50**, 382.

Karel, M. (1976). In Intermediate Moisture Foods (Davies, R., Birch, G.G. & Parker, K.J., eds), p. 4. Applied Science Publishers, London.

Kester, J.J. & Fennema, O.R. (1986 a). Evaluation of an Edible, Heat-Sensitive Cellulose Ether-Lipid Film as a Barrier to Moisture Transmission. Presented at 46th Annual Meeting, Inst. of Food Technologists, Dallas, Tex., June 15-18.

Kester, J.J. & Fennema, O.R. (1986). Food Technol. **40**, 47.

Kroger, M. & Igoe, R.S. (1971). Food Prod. Dev. **5**, 75.

Leistner, L. (1985). In Properties of Water in Foods in Relation to Quality and Stability (Simatos, D. & Multon, J.L., eds), p. 309. Martinus Nijhoff, Dordrecht.

Leistner, L., Rodel, W. & Krispien, K. (1981). In Water Activity : Influences on Food Quality (Rockland, L.B. & Stewart, G.F., eds), p. 855. Academic Press, New York.

Morgan, B.H. (1971). Food Prod. Dev. **5**, 75.

Prince, L.M. (1960). Soap chem. Special. **35**, 103.

Richard-Molard, D. & Lesage, L. (1986). Food Packaging and Preservation (Mathlouthi, M., ed.), p. 165. Elsevier Applied Science Publishers, London.

Richard-Molard, D., Lesage, L. & Cahagnier, B. (1985). In Properties of Water in Foods in Relation to Quality and Stability (Simatos, D. and Multon, J.L., eds), p. 273. Martinus Nijhoff, Dordrecht.

Ricque, A. (1985). Realisation de Films Proteiques Barrieres a l'Oxygene et la Vapeur d'Eau. DEA Science Alimentaires. Universite des Sciences, Montpellier, France.

Rojas de Gante, C. (1985). Mise au point de couches minces comestibles concentrees en alpha tocopherol. DEA Sciences Alimentaires. Universite des Sciences, Montpellier, France.

Schultz, T.H. Owens, H.S. & Maclay, W.D. (1948). J. Colloid Sci. **3**, 53.

Torres, J.A. & Karel, M. (1985). J. Food Proc. Pres. **9**, 107.

Torres, J.A., Motoki, M. & Karel, M. (1985 a). J. Food Proc. Pres. **9**, 75.

Torres, J.A., Bonzas, J.O. & Karel, M. (1985 b). J. Food Proc. Pres. **9**, 93

Troller, J.A. & Christian, J.H. (1978). Water Activity and Food. Academic Press, New York.

Lamb, J.C. & Kinnunen, H.S. (1989). J. Food Sci., 53, [illegible]

Kane, J.H. (1983). In Intermediate Moisture Foods (Davies, R., Birch, G.G. & Parker, K.J., eds), [illegible] Applied Science Publishers, London.

Kester, J.J. & Fennema, O.R. (1986). [illegible]

Kester, J.J. & Fennema, O.R. (1986). Food Technol. 40, [illegible]

Kobayashi, [illegible] (1971). [illegible]

[illegible] (1975). In [illegible] Water Relations of Foods (Duckworth, R.B., ed.), [illegible] Academic Press, London.

[illegible] Influences on Food Quality (Rockland, L.B. & Stewart, G.F., eds), [illegible] Academic Press, New York.

[illegible] (1977). Food Chem., 2, [illegible]

[illegible] (1988). [illegible]

[illegible] (1985). [illegible]

[illegible] (1985). [illegible] Applied Science Publishers, London.

[illegible] (1985). [illegible]

[illegible] (1988). [illegible]

[illegible] (1988). [illegible] Press, [illegible]

[illegible] (1985). [illegible] Press, [illegible]

[illegible] (1979). [illegible]

THE EFFECT OF SULPHITES ON THE DRYING OF FRUIT LEATHERS

R.J. STEELE
CSIRO Division of Food Research
P.O. Box 52, North Ryde 2113
Australia

INTRODUCTION

The drying of fruit to make rolls or leathers offers a convenient method of marketing fruit that would be unacceptable for the fresh fruit market. Liu & Ma (1983) reported on the postharvest loss of selected fruit and vegetables in Taiwan. As shown in Table 1, losses of papaya, for example, were greater than 21% and these losses emphasise the need to develop processes which can prevent wastage of tropical fruit. Fruits grown in tropical regions present special problems in their postharvest handling. At elevated temperatures and humidity, commonly found in the tropics, postharvest spoilage of fruit by moulds is aggravated. Transport of fruit at low temperatures is often not possible and even if it were many tropical fruits are susceptible to chilling injury. Many of the causes of postharvest spoilage can be eliminated by control of the water activity (a_w) of the fruit tissue. Below an a_w of 0.62 moulds, yeasts and bacteria cannot grow and spoil the product. Furthermore, enzyme-mediated reactions at these low water activities proceed at a negligible rate.

Production of fruit with low water activity is usually accomplished by drying. Fruit drying has been used extensively in Australia to preserve apples, apricots, bananas, grapes, peaches, pears and prunes. Many of these fruits are sun-dried and have been successfully exported for over fifty years. Most fruit are sun-dried in piece form and the production of fruit leathers is presently non-existent. In contrast, a steady and growing market for fruit leathers exists in Canada and the United States of America (Anon, 1984). In Canada, the main product is based on apple puree (Moyls, 1981) and the market for these leathers is strong (Moyls, 1987 Personal Communication).

Leathers are made by drying pureed fruit. Puree is usually screeded to a thickness of about 4.5 mm on a tray and dried until a pleasant chewy consistency develops. The resulting product is light and, for many fruits, tasty. A US patent (Morley & Sharma, 1986) for the production of high fibre fruit leathers involves mixing high fructose corn syrup with pectin, locust bean and xanthan gums. A small amount of dehydrated apple flake or frozen orange juice concentrate is added for apple flavoured or

TABLE 1
Postharvest losses in selected fruits and vegetables at the transport, wholesale, and retail levels in Taiwan (Liu & Ma, 1983).

Commodity	% loss at each level			Total
	Transport	Wholesale	Retail	
Chinese cabbage	4.4	22.8	4.9	29.7
Cabbage	4.3	20.9	4.9	28.0
Turnip	2.4	9.5	4.1	15.3
Eggplant	3.0	3.2	1.9	7.9
Cucumber	3.1	3.1	1.5	7.5
Green bean	3.4	0.6	0.2	4.2
Green pepper	2.9	2.8	1.4	6.9
Cauliflower	3.5	5.4	3.1	11.6
Celery	2.5	5.6	2.6	10.4
Tomato	0.6	5.1	2.0	7.6
Ponkan	2.6	1.7	5.0	9.0
Liu-cheng	1.7	3.0	4.3	8.7
Watermelon	10.9	1.2	0.1	12.1
Muskmelon	2.1	5.1	9.3	15.7
Papaya	2.1	7.3	14.3	21.3
Carambola	2.4	6.5	7.2	15.2
Apple	1.6	0.9	3.2	5.7
Banana	-	2.7	6.6	9.0

orange flavoured fruit leathers, respectively. Leathers made from this patent involve dehydration at about 50°C for 1 h compared with over 80°C for 4 h for apple-based leathers in Canada (Moyls, 1981). The dietary fibre content of fruit-flavoured leathers described in the US patent is estimated to be about 25%. This compares favourably with many foods commonly considered to be high in fibre. Bran-based cereals contain about 7% dietary fibre when analysed on an as-consumed basis, i.e. with milk. However, the dietary fibre content of leathers made from whole fruit is excellent. The dietary fibre content of Australian apples, pears, guava, papaya and mango have been determined (Wills & El-Ghetany, 1986; Wills et al., 1986). Based on these analyses, fruit leathers at a moisture content of 15% should contain 40, 20, 19 and 8% dietary fibre for leathers made from guava, apple, papaya and mango, respectively.

Tropical fruits are not used in the US or Canada as a base for the production of fruit leathers. However, Rao & Roy (1980) have satisfactorily produced mango fruit leathers and Chan & Cavaletto (1978) have produced a dried fruit leather made from papaya puree. The shelf-life of leathers is limited by chemical reactions known as non-enzymic browning which discolour the food and can give rise to off-flavours. An essential ingredient to both the mango leather and the papaya leather was the addition of sulphite to prevent discoloration and the production of off-flavours.

Chan & Cavaletto (1978) found that papaya puree with high concentrations of sodium bisulphite (1105 mg SO_2/kg) showed slightly lower rates of drying than papaya puree without sulphite or with a low concentration of sulphite (552 mg SO_2/kg). The lower rate of drying was attributed to a lowering of the vapour pressure of water by the addition of sodium bisulphite. Lowering of the vapour pressure of water by the addition of such a small amount of sodium bisulphite is expected to be small (Lantzke, et al., 1973) and raises the question of what caused the slower rate of drying. Therefore, a study was initiated into the effect of added sulphite on the rate of drying of papaya puree.

MATERIALS AND METHODS

Frozen papaya puree was obtained from Tweed Valley Fruit Processors Pty Ltd in 20 kg cartons. The puree had been sieved to remove seeds and skin, and pasteurised to inactivate enzymes. The puree was stored at -20°C until required and thawed in a cool room for 24 h prior to use.

The experimental plan is shown in Fig. 1. About 7 kg of puree was thawed and sugar (10%, w/w) added and thoroughly mixed. The puree was then divided into seven treatments: a control; 1000 mg SO_2/kg and 2000 mg SO_2/kg added as sodium disulphite (laboratory reagent, Ajax Chemicals Pty Ltd); 1000 mg SO_2/kg and 2000 mg SO_2/kg added as potassium disulphite (laboratory reagent, Ajax Chemicals Pty Ltd); 1000 mg SO_2/kg and 2000 mg SO_2/kg added as sulphur dioxide. Sulphur dioxide was obtained by passing dry gas into a glass trap cooled to -18°C in an ice/sodium chloride slurry. The liquid that condensed was weighed and added to the purees in polyethylene bags which were then quickly sealed. About 0.95 kg of each puree was evenly spread over an aluminium tray lined with polyvinylchloride film. The trays were 0.375 x 0.602 m which corresponded to a loading of 4.2 kg m^{-2} and an initial thickness of about 4 mm. The aluminium trays were placed in a dehydrator and dried at 80°C with an air speed over the trays of 3 m s^{-1}. The mass of each tray was recorded at 15 min intervals.

Moisture was determined by placing about 10 g of leather into a tared aluminium weighing dish containing sand and a small glass stirring rod. The sample, moistened with about 5 ml of 95% ethanol, was thoroughly mixed and dried at 60°C for 2 h. The sample was further dried at 60°C under vacuum (< 100 Pa) for 24 h and reweighed. Concentration of sulphite in purees and leathers was determined by the modified Monier Williams method (AOAC, 1984).

Absorbance of pigment extracted into 70% methanol after homogenization was measured at 400 nm. Absorbance of such alcoholic extracts has been shown to be a measure of the extent of non-enzymic browning in many food products (Nury & Brekke, 1963).

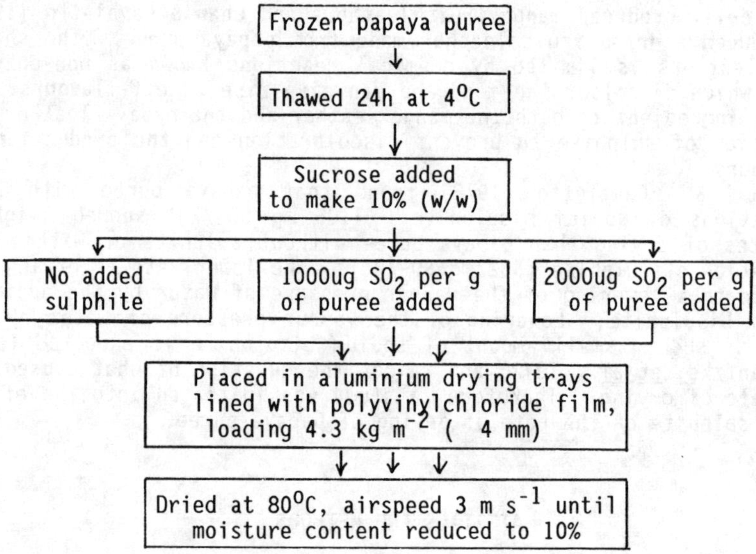

Figure 1. Experimental Plan.

RESULTS AND DISCUSSION

Figs 2, 3, & 4 show that the papaya puree with no added sulphite dried at a faster rate than papaya with added sulphite. The papaya puree with added sulphite dried at a slower rate irrespective of the form of added sulphite.

The times required to reduce the moisture content of the purees to 13% were about 160, 180 and 190 min for the purees with 0, 1000 and 2000 mg of added sulphite per kg of puree, respectively. These drying times were shorter than those found by Chan & Cavaletto (1978) where about 200 min were required at the higher drying temperature of 84°C. Their slower drying rates probably arose from drying in a forced draught oven where the airflow was 0.5 m s^{-1} compared with 3 m s^{-1} in the present study. Moyls (1981) found the drying rate for apple puree increased at higher airflow rates. He estimated the initial drying rate for rates of airflow between 3 and 15 m s^{-1} from the Dalton equation (Eckert & Drake, 1959, p. 470).

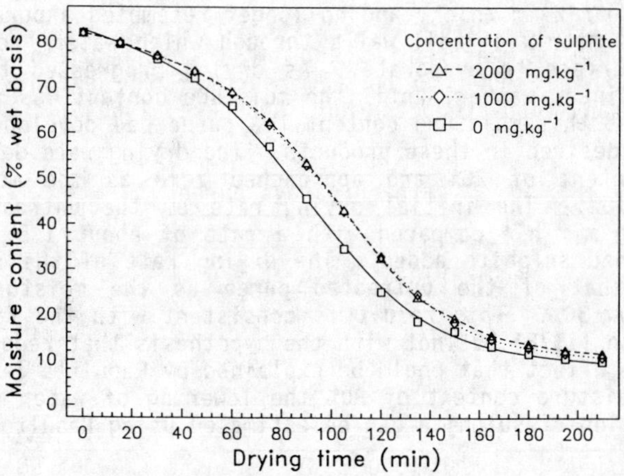

Figure 2. Effect of SO_2 on drying of papaya.

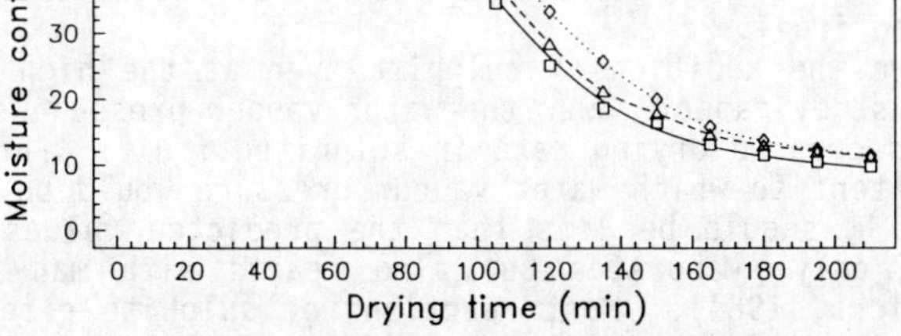

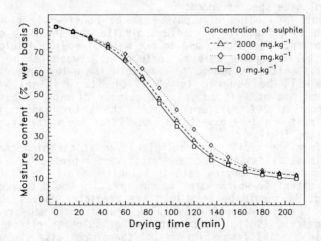

Figure 3. Effect of $K_2S_2O_5$ on papaya drying.

Figs 5, 6, & 7 present the drying rate as a function of moisture content (wet weight basis) and show that the drying rate remained constant, or nearly so, until the moisture content was reduced to about 65%. At this moisture content the puree was less than half its original mass, less than 2 mm thick, and no longer resembled a puree but was more like a solid matrix of cell walls through which water had to diffuse to reach the surface to evaporate. As drying progressed the drying rate fell, in a linear manner, until the moisture content was reduced to less than 20%. At this moisture content the puree had developed the leathery consistency desired in these products. The drying rate decreased below a moisture content of 20% and approached zero as the moisture content approached 10%. The initial drying rate of the untreated purees was about 1.8 kg m^{-2} h^{-1} compared with a rate of about 1.5 kg m^{-2} h^{-1} for puree that had sulphite added. The drying rate of the sulphited puree approached that of the untreated puree as the moisture content was reduced below 30%. This result is consistent with the findings of Chan and Cavaletto (1978) but not with the hypothesis that reduced drying rate was a solute effect that could be explained by Raoult's law.

At a moisture content of 80% the lowering of water vapour pressure by the addition of sulphite can be estimated using Raoult's law:

$$p = p_0 \cdot \gamma \cdot \frac{\text{N water}}{\text{N water} + \text{N sulphite}}$$

where p is the vapour pressure of water after sulphite has been added, p_0 is the vapour pressure of water in the unsulphited puree, γ is the activity coefficient which under ideal conditions is equal to one, N water is the number of moles of water and N sulphite is the number of moles of sulphite species added.

One mole of sodium disulphite can be considered to hydrolyse in solution to form one mole of sodium sulphite and one mole of molecular sulphur dioxide. Therefore, adding one mole of sodium disulphite is the equivalent to adding one mole of sodium ions, two moles of sulphite or three kinetic units. Assuming $\gamma = 1$, then the water vapour pressure of the puree will be lowered to 99.9% of its original value or an insignificant amount. If no loss of sulphur dioxide is assumed during drying, then at 30% moisture content the water vapour pressure of sulphited fruit would be expected to be 99% of the water vapour pressure of unsulphited fruit.

Therefore the addition of sulphite even at the high concentrations used in this study cannot lower the water vapour pressure sufficiently to explain the decreased drying rate in sulphited fruit. Furthermore, even though the extent to which water vapour pressure would be reduced cannot be measured, it should be less than the predicted values above because sulphite not only evaporates but also reacts with many components of fruits (Wedzicha, 1984). Most reactions of sulphite effectively remove the sulphite species from solution and therefore raise the water vapour pressure toward that of the unsulphited puree. Another explanation of the decreased drying rate is that in unsulphited fruit non-enzymic browning reactions effectively remove low molecular weight solutes, such as the reducing sugars in the Maillard reaction to eventually form high molecular weight pigments, and release water in the process (Wedzicha,

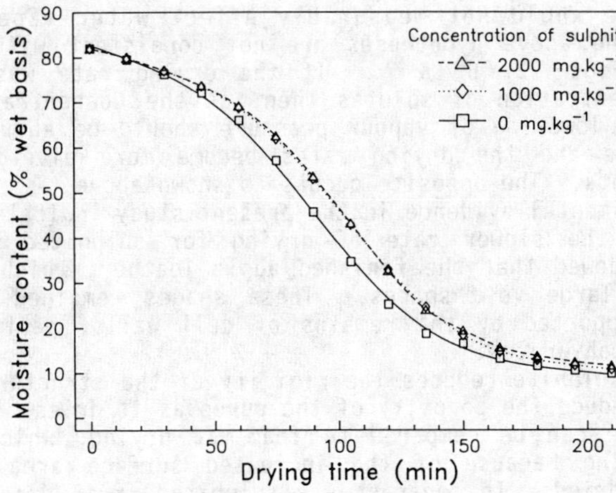

Figure 4. Effect of $Na_2S_2O_5$ on papaya drying.

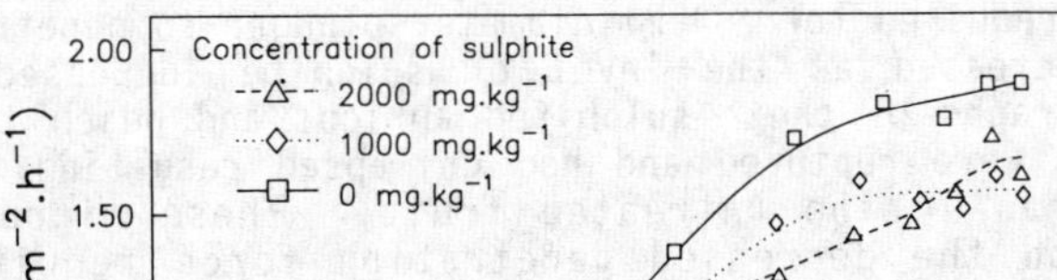

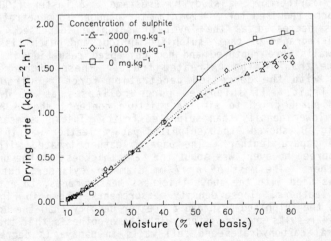

Figure 5. Effect of SO_2 on papaya drying.

1984). Sulphite prevents these reactions and consequently sulphited puree may dry at a slower rate. This explanation is also unlikely because even when non-enzymic browning is extensive the actual loss of reducing sugars is negligible, probably less than 0.5%. A reduction of this magnitude would not measurably affect water vapour pressure. Furthermore, the above hypotheses are not consistent with the results presented in Figs 5, 6, & 7. If the drying rate was affected by increased concentration of solutes then, as the concentration of water decreased, the lowering of vapour pressure should be aggravated in the sulphited puree and the drying rates become more divergent at lower moisture contents. The opposite occurs as shown above.

The experimental evidence in the present study indicates a different mechanism for the slower rate of drying for sulphited papaya puree. Moyls (1981) showed that the finished apple leather, which had no added sulphite, had large void spaces. These spaces remained as the water evaporated, supported by the remains of cell walls, pectins and other structural carbohydrates.

If added sulphite reduces the rigidity of the structural components it will also reduce the porosity of the puree as it dries. Drying of the rigid structure can be compared to foam-mat drying, which is a rapid method of drying because of the increased surface area per mass of product to be dried. In contrast, as sulphited fruit dries, cell walls collapse or partially collapse, and so present a smaller surface area for dehydration. This would result in the sulphited puree having an initial drying rate lower than unsulphited puree. As drying progresses the drying rates of the sulphited and unsulphited purees should converge as drying becomes limited by diffusion of water to the surface of the puree. The hypothesis that sulphite affects the ridigity of cell walls is supported by the findings of Abdelhaq & Labuza (1987) who studied the air-drying of apricots. They showed that, as the level of added sulphite increased, the hardness of unblanched apricots dried to 28% moisture decreased. Furthermore, a study by Prestamo & Fuster (1984) showed that the force required for a 4 mm diameter plunger to penetrate apricots and peaches decreased as the level of sulphite increased. Scanning electron micrographs of their sulphited apricot and peach tissue showed that cell walls were ruptured and had collapsed resulting in a smaller cell volume than in the untreated fruit. These micrographs were consistent with the decreased penetration force required for the sulphited fruit. If sulphited puree collapses as it dries then unsulphited puree dried to similar moisture contents should be thicker and have a lower density than sulphited fruit. Thickness measurements, shown in Fig. 8, showed that sulphited papaya leather was thinner than unsulphited papaya leather. The papaya leather treated with 1000 mg SO_2/kg of puree, however, was about the same thickness as the unsulphited papaya leather. The mass of a 17 mm diameter cylinder, cut from the leathers was used with the above thickness measurements, to estimate the density of leathers. These density measurements, shown in Fig. 9, also showed that the sulphited pulp had a greater density than the unsulphited pulp. These results are consistent with the hypothesis that the rigidity of structural carbohydrates and cell walls in papaya is decreased when the papaya is treated with high concentrations of sulphite; this decrease in rigidity slows the rate of drying.

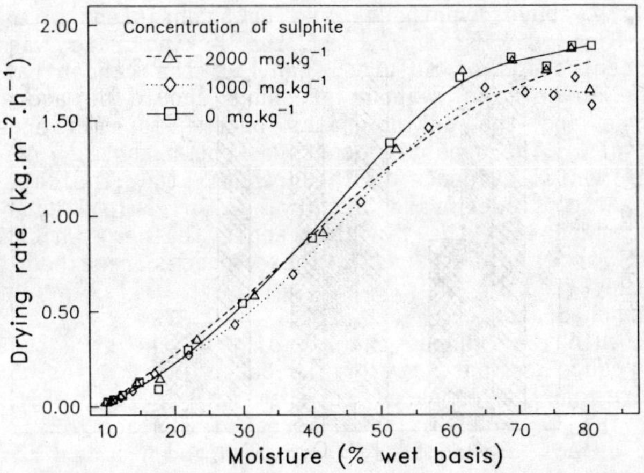

Figure 6. Effect of $K_2S_2O_5$ on papaya drying.

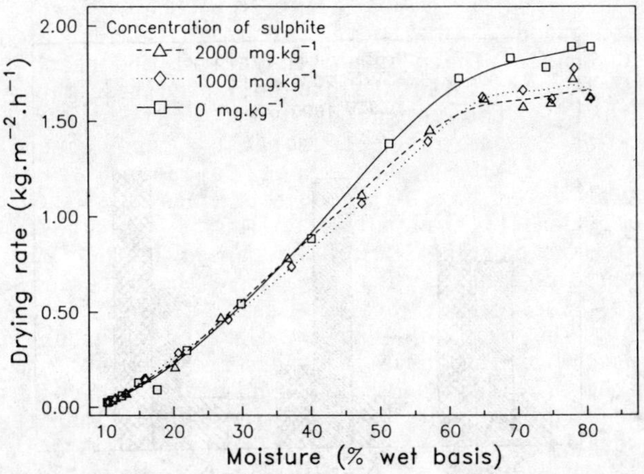

Figure 7. Effect of $Na_2S_2O_5$ on papaya drying.

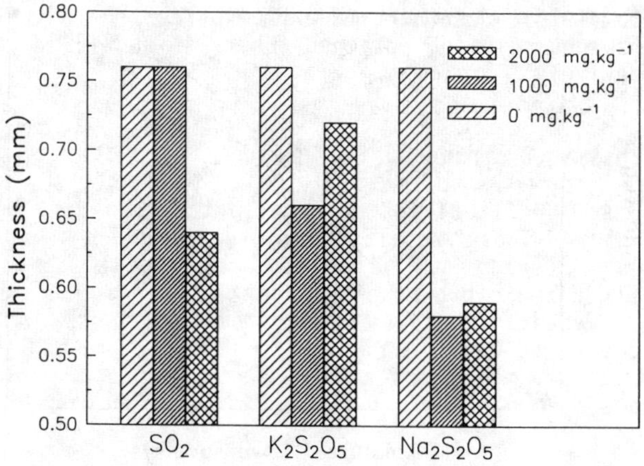

Figure 8. Thickness of papaya leather.

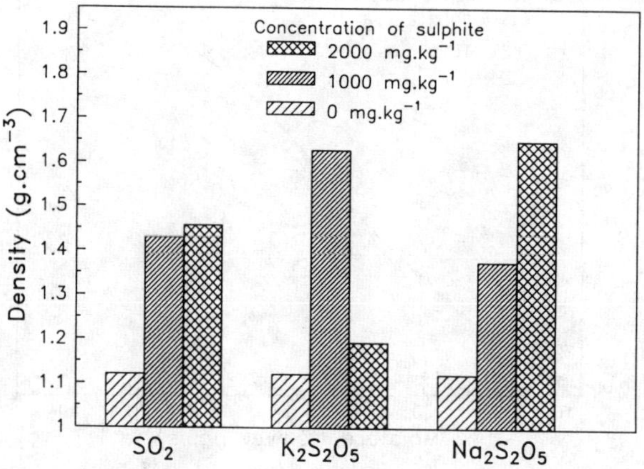

Figure 9. Density of papaya leather.

Retention of Sulphite in Leathers

About 48% of added sulphite was retained in dried leathers where it was added as sodium disulphite. Similar rates of retention were observed by Chan & Cavaletto (1978). Leathers made from puree treated with potassium disulphite and sulphur dioxide retained about 34% and 14% of the original sulphite, respectively.

Extent of Non-Enzymic Browning During Drying

The absorbance, at 400 nm, of 70% methanol extracts of fruit leathers is shown in Fig. 10. These results show that non-enzymic browning is inhibited by the levels of sulphite used in the present study. Inhibition of non-enzymic browning is essential for an acceptable colour in the leather. Canadian leathers, made from apple puree, are sold as a 'health food' and consequently no sulphite is added. Packaging of these leathers requires that the dark colour be hidden and plastics in the package are opaque (Anon, 1984).

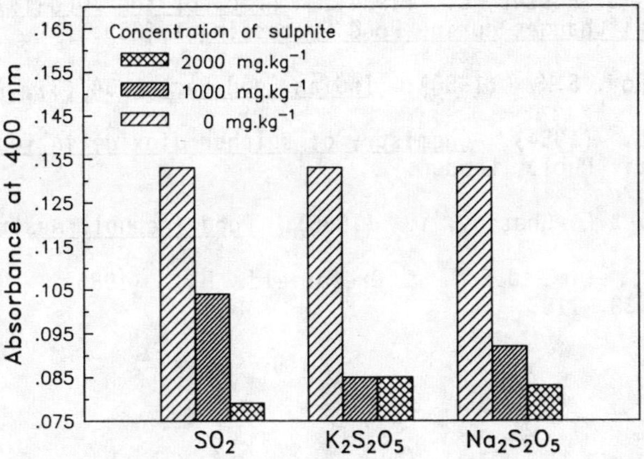

Figure 10. Browning of papaya leather.

REFERENCES

Abdelhaq, E.L. & Labuza, T.P. (1987). J. Food Sci. **52**, 342.

Anonymous. (1984). Food in Canada **44**(10), 24.

Chan, H.T. Jr. & Cavaletto, C.G. (1978). J. Food Sci. **43**, 1723.

Eckert, E.R.G. & Drake, R.M. (1959). Heat and Mass Transfer. McGraw Hill, New York.

Heikal, H.A., El-Sanafiri, N.Y. & Shooman, M.A. (1972). Agric. Res. Rev. **50**, 185.

Lantzke, I.R., Covington, A.K. & Robinson, R.A. (1973). J. Chem. Eng. Data **18**(4), 421.

Liu, M.S. & Ma, D.C. (1983). Postharvest problems of vegetables and fruits in the tropics. Asian Vegetable Research and Development Center, 10th Anniversary Monograph Series, p. 2. Shanhua, Taiwan, Republic of China.

Morley, R.C. & Sharma, C. (1986). Dietary fibre food products and method of manufacture. US Patent **4**, 565, 702.

Moyls, A.L. (1981). J. Food Sci. **46**, 939.

Nury, F.S. & Brekke, J.E. (1963). J. Food Sci. **28**, 95.

Prestamo, G. & Fuster, C. (1984). Proc. of the IUFoST Int. Symposium on Chemical Changes during Food Processing **1**, 269.

Rao, S.V. & Roy, S.K. (1980). Indian Food Packer **34** (3), 72.

Wedzicha, B.L. (1984). Chemistry of sulphur dioxide in foods. Elsevier Applied Sci. Publ., London.

Wills, R.B.H. & El-Ghetany, Y. (1986). Food Technol. Australia **38**, 77.

Wills, R.B.H., Lim, J.S.K. & Greenfield, H. (1986). Food Technol. Australia **38**, 118.

PROBLEMS ASSOCIATED WITH TRADITIONAL MALAYSIAN STARCH-BASED INTERMEDIATE MOISTURE FOODS

C.C. SEOW AND K. THEVAMALAR
Food Technology Division
School of Industrial Technology
Universiti Sains Malaysia
11800 Penang, Malaysia

INTRODUCTION

Many examples abound of well-accepted traditional Malaysian intermediate moisture foods (IMF) which are starch-based. Amongst these are dodol, lempok and wajik. Although these foods do contribute to Malaysian dietary variety, their production technologies are not well-developed enough to ensure their long-term survival especially in the face of stiff competition from large-scale production of Western-style convenience foods. Such small-scale indigenous food technologies must be rapidly upgraded by the application of modern scientific principles to transform, where necessary, the more primitive methods into appropriate technologies. The dearth of detailed scientific studies and documentation where these products are concerned is a contributary factor limiting the scope for improvement of such foods.

This paper discusses the major problems associated with such products in terms of quality and shelf-stability. However, it has to be emphasised at the outset that basic education on personal hygiene and food plant sanitation would go a long way towards reducing many of the problems associated with such products. It is not an over-exaggeration to say that the personal hygiene of many of the food handlers and manufacturers and the cleanliness of the manufacturing premises leave much to be desired. The lackadaisical attitude of some of the local health authorities does not help matters either.

PRODUCTION TECHNOLOGY

In the majority of cases, preparation of traditional IMF involves the mixing or blending of ingredients, cooking with continuous stirring, moulding or extruding, cutting, and packaging. At present, most of these steps involve manual labour when they can be easily mechanised in simple ways.

Ismail & Seow (1982) have previously documented the methods of preparation of such products and only a brief description is given here. Dodol is prepared by mixing and cooking glutinous rice flour, coconut milk, coconut oil and brown sugar until a semi-elastic mass is obtained. Sometimes ordinary rice flour is also added. If the flesh of the durian fruit is included, the higher-priced product that results is called durian dodol. The preparation of lempok or durian "cake" involves mixing of durian flesh with brown sugar and water followed by heat-concentration with continuous stirring. When the desired consistency is attained, the hot mass is moulded and cut into various shapes and sizes. Wheat flour or tapioca starch is an additional ingredient in cheaper grades of lempok. Wajik or glutinous rice "cake" is prepared by first soaking glutinous rice in water for about 4 hours, draining and then steaming at atmospheric pressure until cooked. The steamed rice is then added to a blend of sugar and coconut milk which had been previously heat-concentrated. The mixture is heated with continuous stirring until the desired consistency is achieved. On cooling the mixture sets into a "cake" with most of the individual cooked rice grains being still discernible.

CHEMICAL COMPOSITION

Table 1 shows the chemical composition of the starch-based products mentioned. It is obvious that all of them are low protein-high calorie IMF which may be considered to be excessively sweet by certain people. Another problem is the inconsistent quality as reflected by the relatively large variation in chemical composition of any one particular product. This is a consequence of the variability in the manufacturing process and/or the ingredients used. At present, quality standards for such foods have not been defined, legally or otherwise. Chemical preservatives used include sulphur dioxide and/or benzoic acid. The moisture content and water activity (a_w) of these products are within the intermediate range.

STORAGE STABILITY

Very little scientific work has been carried out to study the stability of traditional Malaysian IMF towards microbial and physico-chemical deterioration. The keeping quality of such foods is primarily dependent on their moisture content-relative humidity equilibria relationships, best illustrated by their sorption isotherms. It is generally accepted that microbial spoilage, non-enzymatic browning, oxidative and hydrolytic rancidity, and degradation of water-soluble nutrients are the deteriorative processes most likely to be predominant over the intermediate moisture/a_w range. This general observation also holds true for traditional Malaysian IMF. More specific to starch-based IMF, however, are the adverse changes in textural characteristics that occur during storage as a result of starch retrogradation.

The partial water sorption isotherms of dodol, lempok and wajik at 30°C over the intermediate a_w range from 0.57 to 0.92, as shown in Fig. 1, were obtained using the isopiestic technique involving equilibration

TABLE 1
Chemical composition of selected traditional Malaysian starch-based IMF

Component	Lempok		Dodol		Wajik
	Mean*	Range	Mean*	Range	Mean**
Water activity	0.74	0.68-0.79	0.74	0.70-0.79	0.95
Moisture, %	20.9	17.6-22.3	16.9	14.3-20.8	40.3
Protein (N x 6.25), %	2.7	1.7-4.1	1.9	0.6-3.1	1.9
Ether extract, %	0.8	0.4-1.4	16.4	12.1-20.6	3.8
Ash, %	1.9	1.3-2.2	1.4	1.0-1.7	0.6
Carbohydrates (by difference), %	73.7	71.3-78.9	63.4	53.8-72.0	53.4
Energy value, kcal/100 g	419	405-550	445	417-493	-
pH	5.4	5.2-5.6	5.1	4.0-5.6	-

* Determined on 5 different brands of samples (in triplicate).
** Determined on a single sample (in triplicate).

of the food over different saturated salt solutions (Rockland, 1960; Greenspan, 1977) in vacuum desiccators. Moisture contents of the equilibrated samples were determined by oven-drying at 105°C for 10 h. Over the intermediate range of a_w, the products contain sufficient water to confer a soft-moist texture which allows them to be consumed "as-is".

Microbial Spoilage

Different types of microorganisms are known to have different approximate lower limits of a_w below which growth is inhibited. With the exception of some halophilic species, bacteria generally require high a_w levels (>0.90) for growth, whilst yeasts and moulds are generally more osmotolerant (Scott, 1957; Leistner & Rodel, 1976). Over the a_w range (0.60 - 0.90) typical of IMF, xerophilic fungi appear to be the principal spoilage agents (Pitt, 1975).

There is a paucity of information on the microbiology of traditional Malaysian starch-based IMF. Microbiological examination of <u>lempok</u>, carried out on freshly prepared samples and after 30 days' storage at 30°C, showed that the majority of the microorganisms present were osmophilic yeasts (Ismail & Seow, 1982). The yeasts probably originated

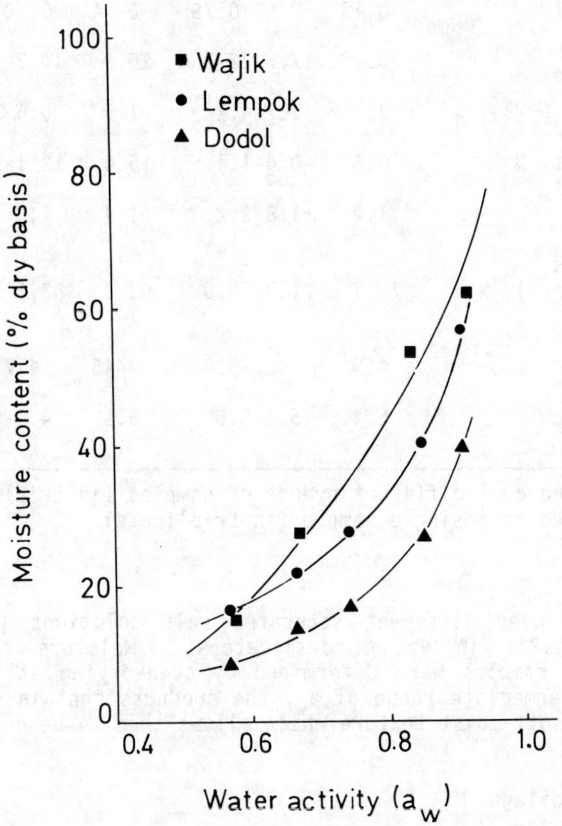

Figure 1. Water sorption isotherms of dodol, lempok and wajik at 30°C.

from post-processing contamination since their vegetative cells, with reported D values of the order of 2 min or less at 65°C, are very easily killed by heat (Gibson, 1973; Corry, 1975) while their ascospores are only slightly more heat resistant (Tilbury, 1976). From a count of 5.7 x 10^2 CFU/g, the number of osmophilic yeasts was observed to increase to 4.3 x 10^4 CFU/g after 30 days' storage at 30°C. These yeasts could, therefore, contribute to spoilage of lempok on prolonged storage.

Since the a_w of lempok and dodol seldom, if ever, exceeds 0.80, it is unlikely that *Staphylococcus aureus*, which appears to exhibit greatest tolerance towards a reduced a_w environment amongst the food-poisoning bacteria, would pose any danger. However, wajik with its relatively high a_w could probably support the growth of certain spoilage and even food-poisoning bacteria.

In the case of traditional IMF such as those mentioned here, other means of enhancing microbiological stability besides the control of a_w must be resorted to, especially where textural or palatability considerations make it undesirable to attain an a_w low enough to prevent spoilage by osmotolerant yeasts and moulds. These include the use of suitable antimycotic agents and packaging materials. Storage of starch-based IMF at refrigeration temperatures for an extended period of time is not recommended since this could accelerate the retrogradation process, thereby leading to undesirable texture changes. The observance of the most stringent hygienic practices at all stages in the preparation of these products cannot be over-emphasised.

Starch Retrogradation and Texture

One of the major problems encountered with traditional starch-based IMF is the hardening of the products during storage, even in the absence of water loss. Texture is an important quality factor governing the acceptability of such ready-to-eat semi-moist foods. As starch is a major component of these products, it is to be expected that any changes in the gelatinised starch fraction that occur during storage will affect their textural properties and hence their acceptability and shelf-life.

It is a well-known fact that progressive association of starch molecules during storage or ageing of gelatinised starch, a phenomenon termed as retrogradation, results in an increase in consistency and spontaneous precipitation of starch from solution or, in the case of gels, increased firmness and finally shrinkage and syneresis (Collison, 1968). Retrogradation is basically a spontaneous crystallization process with the system moving towards a thermodynamically more stable condition. The branched amylopectin molecules are less apt to retrograde than the linear amylose molecules (Collison, 1968). There appears to be evidence to suggest that amylopectin retrogradation is reversible by the application of heat whilst that of amylose is not (Eliasson, 1985).

Starch retrogradation is affected by many factors, amongst them the water content of the system. There is, unfortunately, little information that directly relates retrogradation and texture to moisture content or a_w in starch-based IMF. Some experiments were thus carried out to study the effects of storage and moisture content/a_w on the texture profile parameters of dodol and wajik. The texture profile analysis was conducted using the Instron Universal Testing Instrument (Model 1140) following the general procedures outlined by Bourne (1968, 1978).

Regular pieces of the products, measuring 15 x 15 x 20 mm, were compressed between the stationary horizontal sample support plate of the machine and a moving horizontal perspex sheet (60 x 60 x 3 mm) mounted on the end of a standard compression anvil. Excessive adhesiveness was evident if the compression anvil was used without the perspex plate. A crosshead speed of 100 mm/min and a chart speed of 200 mm/min were used in the compression of the pieces from 20 mm to 5 mm (i.e. 75% compression). Two compressions were made on each sample to give a "double-bite" profile (Bourne, 1968). The texture profile parameters were then determined from the force-distance curves obtained. At least five replicate tests were conducted for any one treatment and the mean value for each textural parameter was then calculated. The area under a curve was determined using a planimeter. All Instron measurements were made at room temperature.

Changes in textural parameters during storage : To study the effects of storage on texture profile parameters of *dodol* and *wajik*, sample pieces were sealed in tinplate cans and stored at the desired temperatures. Samples were withdrawn at appropriate intervals of time and the textural parameters determined at room temperature.

Fig. 2 shows typical traces of the force-distance curves obtained during two compressions of *dodol* samples with a moisture content of 16.1% (dry basis) and 0.74 a_w stored at 30°C for various times. When freshly prepared, *dodol* exhibits the following characteristics :

(a) High cohesiveness (defined as the ratio of the second bite area to the first bite area)
(b) High springiness or elasticity (defined as the height that the product recovers after the first compression)
(c) Little or no adhesiveness (defined as the negative force area for the first bite)
(d) No fracturability or brittleness. However, a distinct fracturability peak appears in the first compression curve on storage.

Typical Instron traces obtained with *wajik* at 67.6% moisture (dry basis) and 0.95 a_w stored at 30°C and 5°C are shown in Figs 3 and 4, respectively. At 30°C, the low initial slopes of the curves indicate high deformability with little or no fracturability. Since the area of the second compression curve is much less than that under the first compression curve, it may be deduced that *wajik* exhibits low cohesiveness. Adhesiveness appears to be a more pronounced characteristic for *wajik* than for *dodol* under the conditions of the test. Little change in the characteristics of the curves obtained occurred during storage at 30°C. In contrast, samples of *wajik* stored at 5°C became slightly harder, less adhesive and more brittle, as evidenced by the emergence of a definite fracturability peak in the curve of the first bite.

A more detailed analysis of the effects of storage on the individual texture profile parameters are shown in Figs 5-10. An apparent increase in hardness of *dodol* with time was observed (Fig. 5), probably as a result of starch retrogradation. On the other hand, *wajik* appeared to soften initially before hardening. The reason for this is as yet unclear. What is apparent, however, is that the storage time taken for a reversal from softening to hardening to occur is considerably shortened at the lower temperature of storage. Wajik samples stored at 5°C were definitely harder than the corresponding samples stored at 30°C. Such an observation is consistent with what is already known about the starch retrogradation phenomenon. Above freezing temperatures, it is an established fact that the lower the storage temperature, the faster the rate of retrogradation (Kalb & Sterling, 1962; Collison, 1968; Nakazawa et al., 1984; Eliasson, 1985). Very little change in cohesiveness of the products under study occurred during storage. Figs 6-8 show that dodol exhibits higher gumminess, chewiness and springiness than wajik. However, while these parameters decrease sharply with time where dodol is concerned, they were better maintained in wajik. Adhesiveness of wajik also declined with time, the rate and extent of decline being greater at the lower storage temperature (Fig. 9).

Dodol appeared to keep well for the first few days at 30°C without any sign of fracturability on compression, after which a definite fracturability peak appeared (Figs 2 and 10). On the other hand, wajik exhibited no fracturability at 30°C, even after 16 days' storage. At 5°C, however, the individual cooked rice grains fell apart easily on compression, thus giving rise to a fracturability peak in the curve of the first bite.

The different storage behaviour of dodol and wajik at 30°C can probably be accounted for to a large extent by the presence of different starch fractions in the products. The starch in wajik, being derived from glutinous or waxy rice, is nearly 100% amylopectin which is known to have a greater resistance to retrogradation than amylose. The particular sample of dodol tested was prepared using both ordinary and glutinous rice and thus would have a higher proportion of amylose and a greater susceptibility to retrogradation which, in turn, would lead to increased hardening and fracturability.

If indeed the changes in textural properties during storage were due primarily to starch retrogradation, then it would be expected that, on heating the stored samples, the textural parameters would regain, at least partially, their original values in the fresh state. A comparison of the texture profile parameters of freshly prepared wajik with those of stored samples and stored and heated samples found this to be true (Table 2). Samples were stored at 5°C for 21 days and heating of the stored samples was effected by immersing the cans containing the wajik samples in boiling water for 15 min. It is clear that such a heating treatment is capable of reversing the effect of storage on the textural properties of wajik. Again this is consistent with the fact that retrogradation of amylopectin is heat-reversible.

Influence of a_w on texture profile parameters of wajik: Moisture content or water activity plays a critical role in influencing the textural characteristics of IMF in that unduly high or low moisture levels would result in excessive softness or hardness, respectively. In other words, there is generally a critical range of moisture content or a_w above and below which the product would be deemed less than satisfactory from the textural viewpoint. Unfortunately, there is a dearth of information on the effects of a_w on the texture profile parameters of ready-to-eat IMF.

Studies on the relationships between the thermodynamics of water vapour sorption and textural properties have been confined to only a few food materials, particularly pre-cooked freeze-dried beef (Kapsalis et al., 1970a; Kapsalis et al., 1970b; Reidy & Heldman, 1972; Kapsalis &

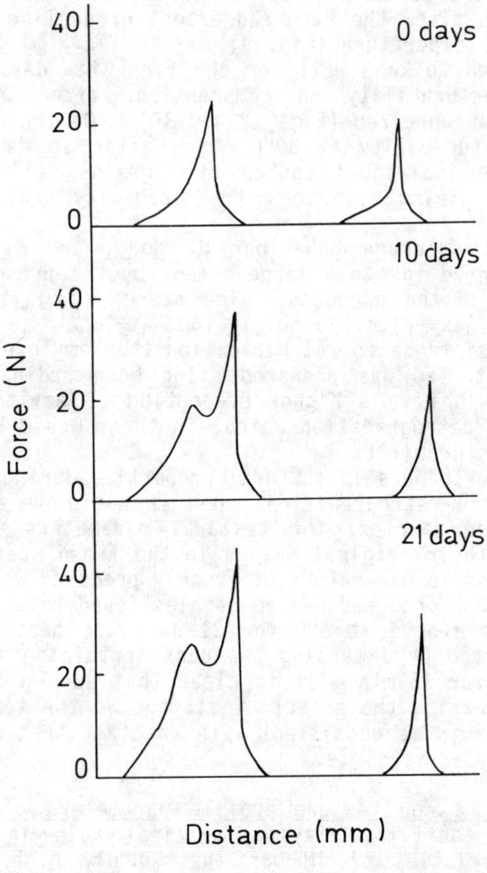

Figure 2. Instron compression profiles of <u>dodol</u> (0.74 a_w) stored at 30°C.

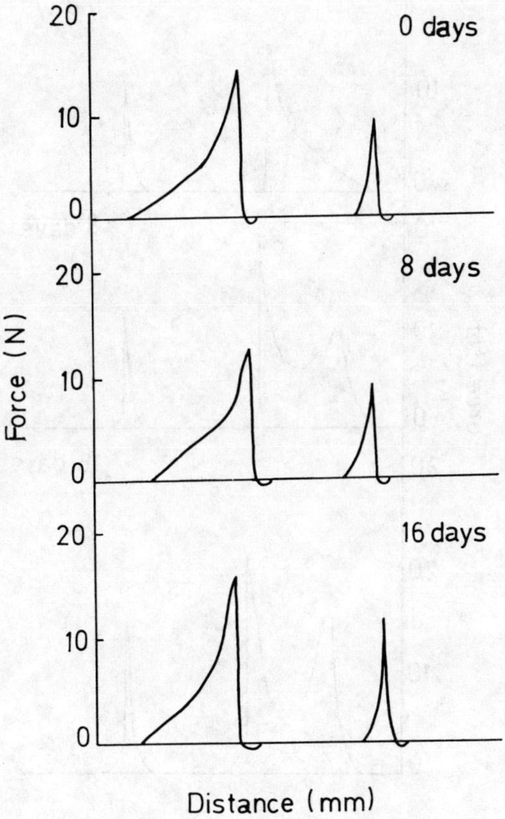

Figure 3. Instron compression profiles of wajik (0.95 a_w) stored at 30°C.

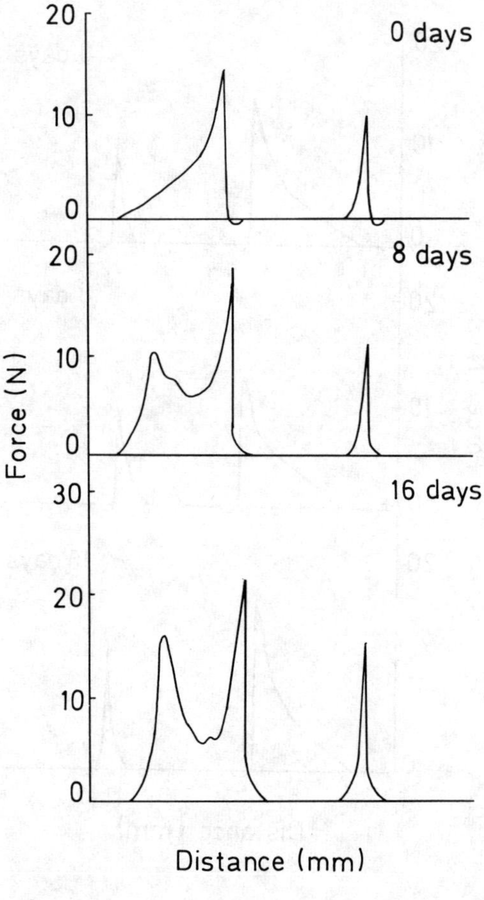

Figure 4. Instron compression profiles of wajik (0.95 a_w) stored at 5^oC.

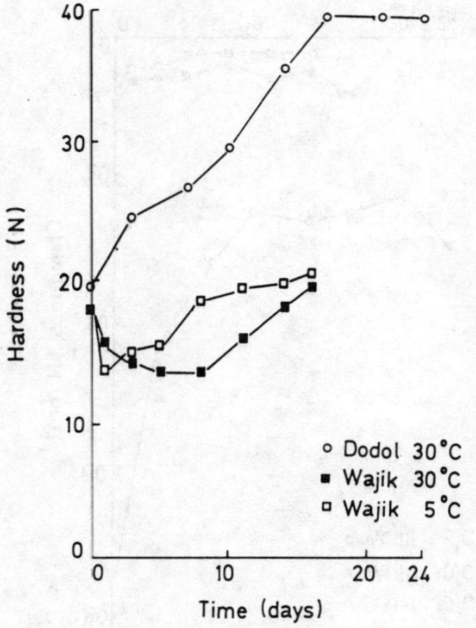

Figure 5. Effects of storage on hardness of dodol and wajik.

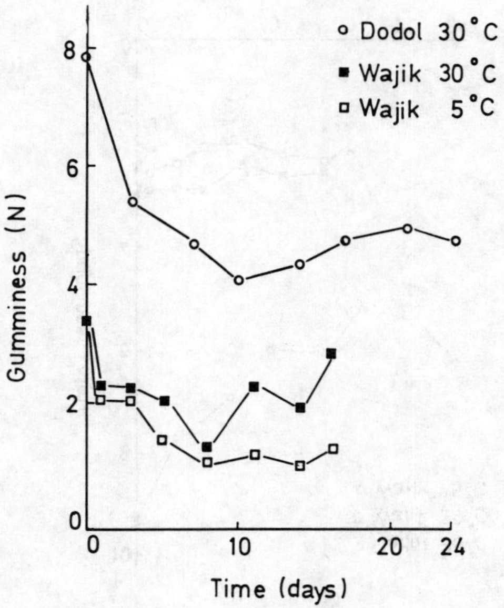

Figure 6. Effects of storage on gumminess of dodol and wajik.

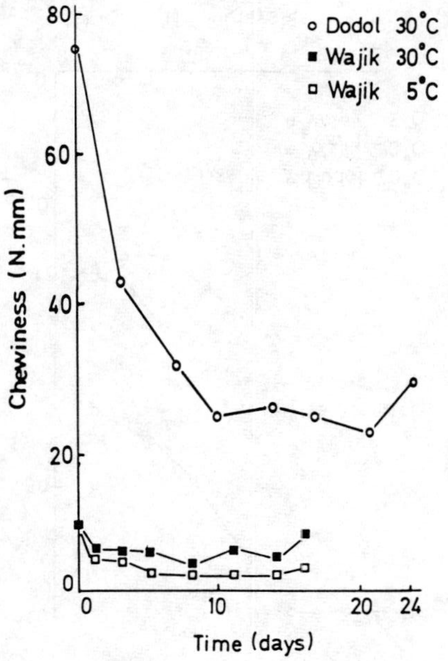

Figure 7. Effects of storage on chewiness of *dodol* and *wajik*.

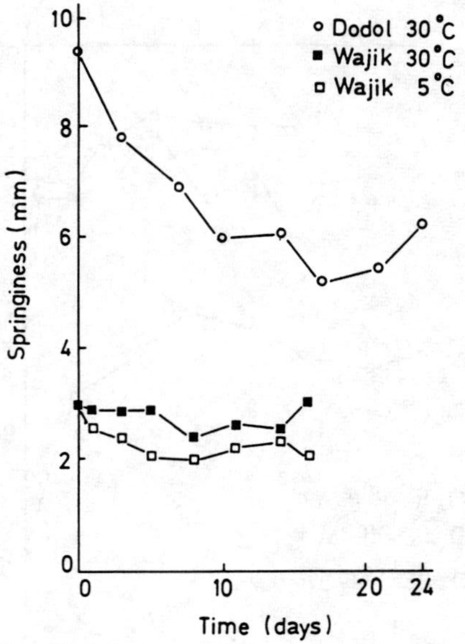

Figure 8. Effects of storage on springiness of *dodol* and *wajik*.

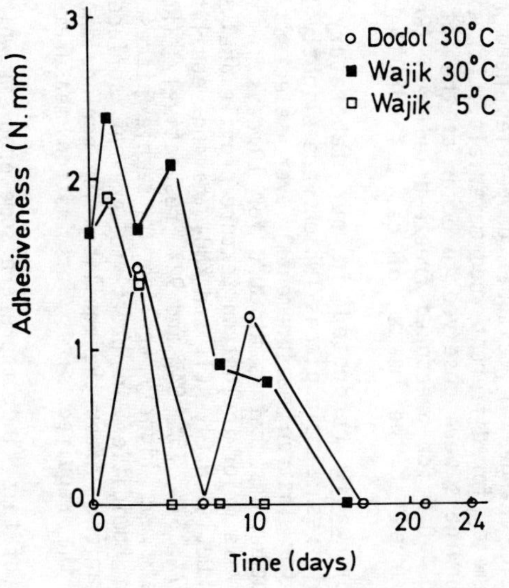

Figure 9. Effects of storage on adhesiveness of dodol and wajik.

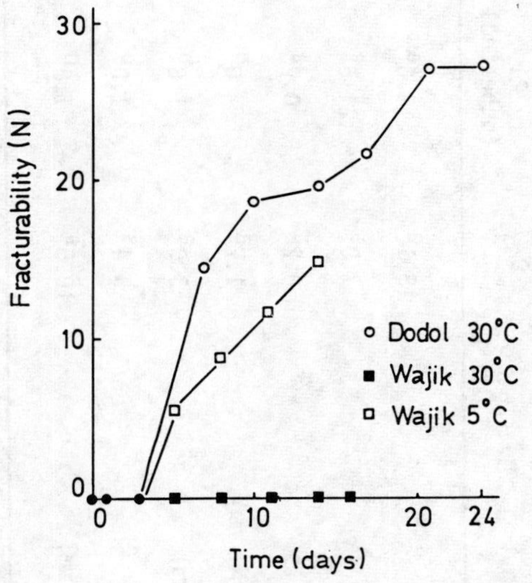

Figure 10. Effects of storage on fracturability of dodol and wajik.

TABLE 2
Effects of heating on texture profile parameters of *wajik*

Textural parameter	Storage time at 5°C (days)		
	0	21 (unheated)	21 (heated)
Hardness (N)	18.1[a]	19.6[a]	17.6[a]
Fracturability (N)	0[a]	21.6[a]	0[a]
Cohesiveness	0.2[a]	0.2[a]	0.2[a]
Adhesiveness (N. mm)	1.7[a]	0[b]	2.5[a]
Springiness (mm)	3.0[a]	1.8[b]	2.9[a]
Gumminess (N)	3.4[a]	1.0[b]	2.9[a]
Chewiness (N. mm)	10.5[a]	1.8[b]	7.8[a]

For each textural parameter, figures followed by the same letter are not significantly different at the 5% probability level.

Segars, 1976). Work done prior to 1975 has been reviewed by Kapsalis (1975). More recently, the instrumental texture profile of apple tissue over the full a_w range has been reported (Bourne, 1986). These studies show that a_w exerts a profound influence on the textural properties of foods, particularly at critical levels such as the BET monolayer level. However, dramatic changes in textural properties need not necessarily occur at the same a_w for different food materials. For example, maximum hardness and chewiness were observed to occur at the intermediate water activities of 0.4-0.6 for precooked freeze-dried beef (Reidy & Heldman, 1972), but occurred at the low a_w of ca. 0.12 in the case of apple (Bourne, 1986).

Studies on *wajik* conducted in our laboratory involved the equilibration of fresh sample pieces (which were slightly larger than the required size for Instron measurements) over selected saturated salt solutions in vacuum desiccators at 30°C for 10 days in order to attain different a_w levels prior to Instron texture profile analysis. The salts used were NaBr, $CuCl_2$, NaCl, KCl, and KNO_3 giving equilibrium relative humidities of 57%, 67%, 75%, 84% and 92%, respectively (Rockland, 1960). Potassium sorbate was lightly applied to sample pieces stored at the higher relative humidities in order to prevent surface mould growth during equilibration. After the equilibration period, the sample pieces were cut into the required sizes for Instron measurement as earlier described.

In studies of this kind, there are some problems which are inherent in the method used in the conditioning of a sample to arrive at a desired

a_w. For example, microbial (especially fungal) growth may occur at the high relative humidity levels. Where the present investigation is concerned, it cannot be denied that some textural changes are bound to occur during the relatively long equilibration period. As such, the Instron textural parameters measured do not simply reflect the effects of a_w *per se* but also changes that might have taken place during the equilibration process. Nevertheless, such "storage" changes are themselves very likely to be influenced to a large extent by a_w. Overall, therefore, the Instron measurements obtained on samples equilibrated at different relative humidities should in general still be relective of the influence of a_w on texture and the complication just described should not detract much from the validity of the results obtained and the conclusions drawn therefrom.

The Instron compression curves of samples at a_w levels of 0.92, 0.84 and 0.75 were found to be qualitatively similar. These samples exhibited low cohesiveness, high springiness, slight adhesiveness, high deformability and no fracturability or brittleness. However, dramatic changes in the characteristics of the Instron curves were obtained as a_w was reduced to 0.67, the most interesting being the emergence of a definite fracturability peak in the curve of the first bite, the loss of adhesiveness, and the sharp increase in hardness. Such changes were further accentuated for samples at 0.57 a_w (results not shown). At 0.67 a_w and below, the cooked rice grains were observed to crumble easily on compression, thus giving rise to a fracturability peak.

The effects of a_w on the individual texture profile parameters were analysed in greater detail and are shown in Figs 11-13. With the exception of adhesiveness and cohesiveness, the other texture profile parameters were found to increase as a_w was reduced from 0.92 to 0.67. Water activity did not appear to have any effect on cohesiveness of the product over the range of a_w studied. As shown in Fig. 11, there was no evidence of fracturability in samples at a_w levels higher than 0.75 (corresponding to a moisture content of ca. 45%). However, fracturability appeared quite suddenly on reduction of a_w to 0.67. The mechanical parameters of hardness, chewiness, and gumminess were also observed to increase steeply as a_w was decreased below 0.75 (Figs 11 & 12), whilst adhesiveness disappeared at and below the same a_w level (Fig. 13). However, as shown in Fig. 13, there was a progressive increase in springiness with decreasing a_w, with no evidence of any dramatic changes or discontinuity.

It would appear that *wajik* exhibits a critical moisture content or a_w below which most of its mechanical textural properties are drastically (and adversely) affected. The occurrence of a critical a_w at the relatively high level of ca. 0.75 may be an indication that a transition in the mode or mechanism of water sorption from multilayer adsorption to capillary condensation is associated with major textural changes in the product. On the other hand, any such relationship could be fortuitous. Whilst a reduction in moisture content or a_w would enhance the stability of the product through inhibition of microbial growth, it could, if taken too low, adversely affect the acceptability of the product from a textural standpoint. As such, the water activity of *wajik* should be maintained at a level between 0.80 and 0.90 to ensure good textural properties and, at the same time, afford some protection against microbiological (primarily bacteriological) spoilage.

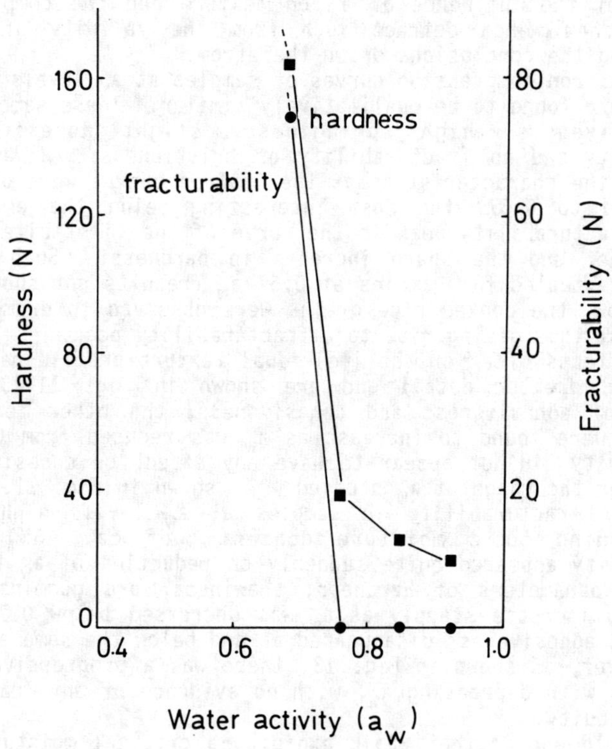

Figure 11. Instron texture profile parameters of hardness and fracturability of wajik as a function of a_w.

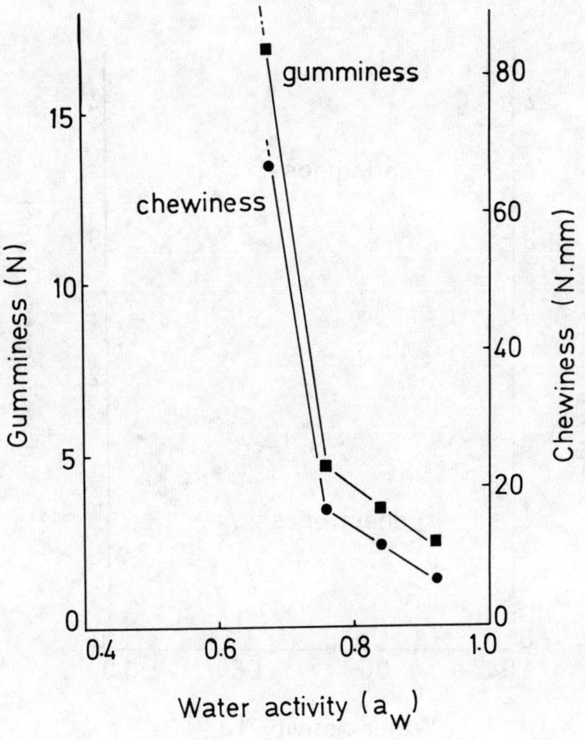

Figure 12. Instron texture profile paramaters of gumminess and chewiness of wajik as a function of a_w.

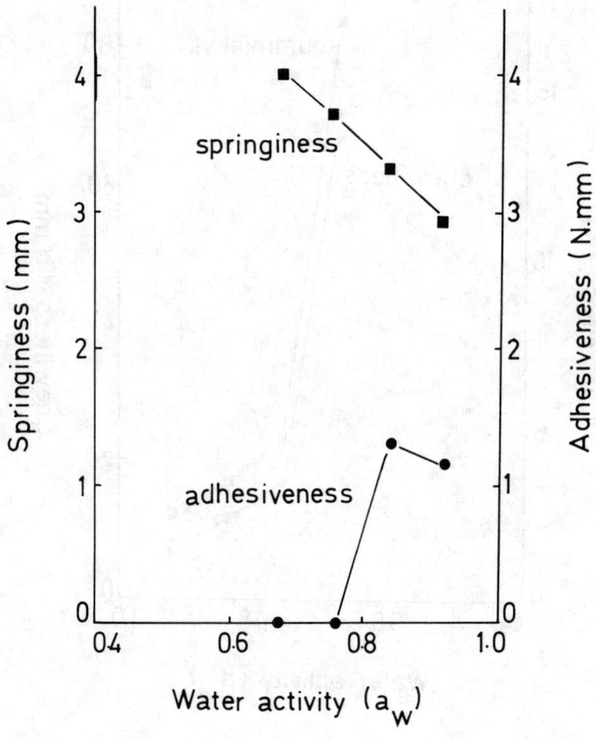

Figure 13. Instron texture profile parameters of adhesiveness and springiness of wajik as a function of a_w.

CONCLUSION

The keeping quality of starch-based IMF such as those mentioned in this paper is determined not only by deteriorative reactions of microbiological or chemical origin but also by adverse textural changes that occur during storage as a result of starch retrogradation. Much remains to be done to overcome such problems. The possibility of using certain mechanical textural properties, in particular fracturability, as indices of quality of such products might be well-worth exploring.

ACKNOWLEDGEMENTS

Financial support, in the form of a short-time research grant, from Universiti Sains Malaysia is gratefully acknowledged.

REFERENCES

Bourne, M.C. (1968). J. Food Sci. **33**, 223.

Bourne, M.C. (1978). Food Technol. **32**(7), 62.

Bourne, M.C. (1986). J. Texture Studies **17**, 331.

Collison, R. (1968). In Starch and its Derivatives, 4th edn. (Radley, J.A. ed.), p. 325. Academic Press, London.

Colwell, K.H., Axford, D.W.E., Chamberlain, N. & Elton, G.A.H. (1969). J. Sci. Food Agric. **20**, 550.

Corry, J.E.L. (1975). In Water Relations of Foods (Duckworth, R.B., ed.), p. 325. Academic Press, London.

Eliasson, A.-C. (1985). In New Approaches to Research on Cereal Carbohydrates (Hill, R.D. & Munck, L., eds), p. 93. Elsevier, Amsterdam.

Gibson, B. (1973). J. appl. Bact. **36**, 365.

Greenspan, L. (1977). J. Res. Nat. Bur. Stand. **81A** (1), 89.

Ismail, N. & Seow, C.C. (1982). In Proc. Int. Symp. on Food Technology in Developing Countries (Berry, S.K., ed.), p. 299. Universiti Pertanian Malaysia, Serdang, Selangor, Malaysia.

Kalb, A.J. & Sterling, C. (1962). J. Food Sci. **26**, 587.

Kapsalis, J.G. (1975). In Water Relations of Foods (Duckworth, R.B., ed.), p. 627. Academic Press, London.

Kapsalis, J.G., Drake, B. & Johanson, B. (1970a). J. Texture Studies 1, 285.

Kapsalis, J.G. & Segars, R.A. (1976). Lebensm. Wiss. u-Technol. **9**, 383.

Kapsalis, J.G., Walker, J.E. & Wolf, M. (1970b). J. Texture Studies **1**, 464.

Leistner, L. & Rodel, W. (1976). In Intermediate Moisture Foods (Davies, R., Birch, G.G. & Parker, K.J., eds), p. 120. Applied Sci. Publ., London.

Nakazawa, F., Noguchi, S., Takahashi, J. & Takada, M. (1984). Agric. Biol. Chem. **48**(1), 201.

Pitt, J.I. (1975). In Water Relations of Foods (Duckworth, R.B., ed.), p. 273. Academic Press, London.

Reidy, G.A. & Heldman, D.R. (1972). J. Texture Studies **3**, 218.

Rockland, L.B. (1960). Anal. Chem. **32**(10), 1375.

Scott, W.J. (1957). Adv. Food Res. **7**, 83.

Tilbury, R.H. (1976). In Intermediate Moisture Foods (Davies, R., Birch, G.G. & Parker, K.J., eds), p. 138. Applied Sci. Publ., London.

EFFECT OF WATER IN VEGETABLE OILS WITH SPECIAL REFERENCE TO PALM OIL

C.L. CHONG AND A.S.H. ONG
Palm Oil Research Institute of Malaysia
No. 6, Persiaran Institusi, Bandar Baru Bangi
P.O. Box 10620, 50720 Kuala Lumpur
Malaysia

INTRODUCTION

Preservation of food is the maintaining of the quality of the food or the prevention of its deterioration. In order to effectively preserve the food, an understanding of the causes of the deterioration would be a great help. For oils and fats, the major deteriorations could broadly be ascribed to two major causes, namely that of hydrolytic and that of oxidative deterioration, both of which together accounting for the production of rancid off-flavour and off-odours.

HYDROLYTIC DETERIORATION

The hydrolysis of the ester bonds of triglycerides is due either to the attack by water or by lipase, resulting in the formation of free fatty acids (ffa) which are responsible for rancid off-flavours, partial glycerides and glycerol, if the deterioration is in an advanced stage.

Vanneck et al. (1951) demonstrated the effect of water content and temperature on the acidification of palm oil. Palm oil samples with water contents of 0.25% and 0.80% were stored at 18°C, 55°C and 75°C. At 18°C after 62 days, there was no appreciable difference between the two samples. At 55°C after 55 days, a noticeable difference in the ffa of the two samples was observed. At 75°C, after only 36 days, the higher moisture sample had an ffa value twice that of the lower moisture sample. This amply demonstrates the involvement of water in the hydrolytic reaction and the effect of its excess in terms of ffa formation. Vanneck et al. (1951) also gave the first indication of the autocatalytic effect of ffa by showing the slow acidification rates of a synthetically prepared oil having a composition similar to that of palm oil and a palm oil sample treated with caustic soda.

In a thesis on the "Spontaneous hydrolysis of glyceridic oils and in particular palm oil", Loncin (1952) established the acidification of vegetables oils on a more theoretical basis albeit with some misconclusions. He confirmed Vanneck et al.'s earlier observation that the presence of water was required for the acidification of the oil, i.e. as an active component in the hydrolytic reaction. His contention that this acidification was purely a chemical reaction and not due to enzymes or micro-organisms was later shown to be wrong due to the fact that in his experiments with added bactericides, the temperature at which the experiments were conducted was too high (> 50°C) for bacterial growth. Certain fungi and bacteria have now been isolated and shown to possess great fat-splitting potency. In fact, some micro-organisms were found to be able to increase the ffa content rapidly. For example, *Geotrichum candidum* was shown to be able to increase the ffa from 6% to 21% in just 21 days, provided the oil contains sufficient water and dirt to enable it to grow (Stork, 1960). It is now generally accepted that both hydrolytic and enzymatic acidification of oils could occur simultaneously under certain conditions, e.g. optimal water content, temperature, dirt level, etc.

One of the more important observations of Loncin (1952) is the fact that the initial concentration of the ffa present in the oil is the most important factor influencing the rate of the hydrolytic reaction, apart from that of water. He expressed the autocatalytic hydrolysis of palm oil in an equation, viz.

$$\frac{da}{dt} = ka \qquad (1)$$

for oils with low ffa content (< 10%) where

- a is the percentage of ffa;
- t is the time period of storage in tens of days; and
- k is the reaction constant.

The reaction constant (k) is temperature-dependent and approximately doubles for every 10°C increase in temperature. It must be noted that this relationship was derived from experimental results whereby the oils under investigation were in constant contact with excess moisture.

The above equation implies that the hydrolytic reaction is influenced by the initial amount of ffa present in the oil, the temperature, and the time of the reaction. The moisture content of the oil is significant only in so far as it renders the reaction possible. If there is no moisture left, the reaction will cease. In theory, the stabilisation of palm oil or any other vegetable oils, in terms of ffa formation, could be achieved by completely dehydrating the oil. In practice, this is difficult to achieve, given the hygroscopic nature of the oil.

Loncin (1952) also showed that low molecular weight fatty acids are more active than long chain fatty acids at a given level (expressed as weight percentages), viz. C_{16} to C_{18} as compared with C_4 to C_{14}. The degree of unsaturation has no effect at all on the catalytic activity of the acids.

Experiments with other vegetable oils showed that the phenomenon of spontaneous autocatalytic hydrolysis is general in nature and is independent of the oil type, all with roughly similar rates. Kinetic studies show that at the initial stage where the ffa level is low (< 10%), the reaction proceeds via the route:

$$\text{Triglycerides} + H_2O \longrightarrow \text{Diglycerides} + \text{ffa}$$

Above 10% ffa but less than 20%, there is some production of monoglycerides:

$$\text{Diglycerides} + H_2O \longrightarrow \text{Monoglycerides} + \text{ffa}$$

Above 20% ffa, some of the monoglycerides are converted to glycerol, viz.

$$\text{Monoglycerides} + H_2O \longrightarrow \text{Glycerol} + \text{ffa}$$

At a level of approximately 30% by weight of ffa, the molar concentration of diglycerides reaches a steady state, i.e. their rate of formation is equal to their rate of hydrolysis into monoglycerides and ffa. The molar concentration of diglycerides to triglycerides at this stage is close to 3/2. This corresponds to an even probability for the hydrolysis of each of the ester groups present. This was confirmed by experiments carried out on pure partial glycerides.

So far no reference has been made to the solubility of water in oils. Loncin (1955) noted that the solubility of water in oil is dependent on the temperature, the amount of ffa and the type of triglycerides present in the oil. In general, water solubility in vegetable oils increases with an increase in temperature and the level of ffa. The solubility is also high for oils with shorter acid chains as shown in Tables 1 and 2.

TABLE 1
Solubility of water in vegetable oils at 60°C

Type of oil	Solubility of water (%)
Coconut oil	0.285
Palm oil	0.230
Soya oil	0.190

TABLE 2
Solubility of water in palm oil at 60°C with different levels of ffa

Free fatty acids (%)	Solubility of water (%)
0.20	0.23
0.25	0.27
6.50	0.32
15.00	0.42

Hidler (1968) showed that the solubility of water in liquid oils and fats is independent of the type of oil or fat considered if the mole fraction concept is used. In general, for a three-phase system comprising of water vapour, liquid water and water dissolved in the oil, the solubility of water in oil can be expressed by the equation :

$$\ln X_s = 2.5 - \frac{1{,}600}{T} \quad (2)$$

where X_s is the mole fraction of water in the liquid phase at the solubility limit and $273K<T<373K$. For a two-phase system of just water in oil and water vapour, the solubility is expressed by :

$$\ln \frac{XP_s}{P} = 2.5 - \frac{1{,}600}{T} \quad (3)$$

where X is the mole fraction of water;
P is the partial vapour pressure at temperature T;
P_s is the saturated vapour pressure at temperature T and $273K<T<373K$.

These equations apply to triglycerides with average molecular weights in the range 650 to 960. Fig. 1 shows the solubility limits for palm oil calculated from Equation 3 with an average molecular weight of 836 (calculated from weighted fatty acid composition contributions). As can be observed, the solubility of water in palm oil is about 0.20% at 50^oC. Fig. 2 shows the correlation between water content and water activity (a_w) at 50^oC. For refined, bleached and deodorised (RBD) palm oil with a water content of about 0.1%, the water activity is about 0.5.

From the above account, it can be seen that the solubility of water in palm oil is low. Since only water in solution or at an oil/water interface can participate in the hydrolytic reaction, the reaction constant k outlined above in Equation 1 is independent of the excess water concentration. If the water content is below the solubility limit of water in oil, then water content or a_w becomes the rate controlling factor and the equation is modified to :

$$\frac{da}{dt} = k\,(a)\,(a_w) \quad (4)$$

The confusing and conflicting results obtained by earlier workers viz the chemical theory of Loncin and the microorganism effects observed by Fickendey (1910), Bunting _et al_. (1934), Vervloet (1953), Desassi (1957) and Coursey (1958) could now be resolved. Under sterile conditions, the chemical hydrolytic rate is not affected by impurities

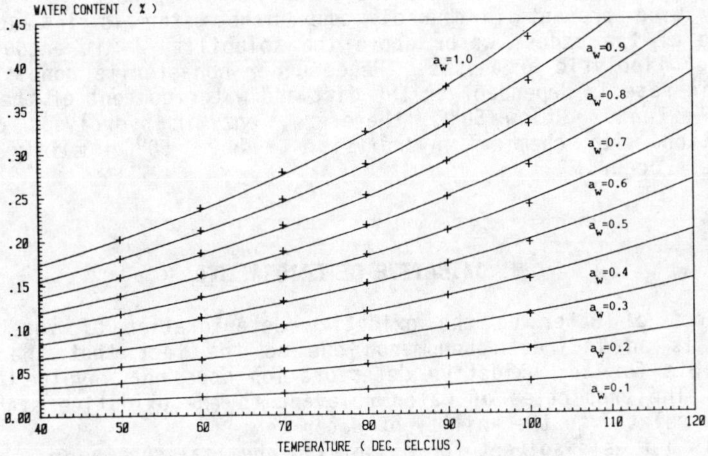

Figure 1. Palm Oil water content vs temperature at different a_w levels (PO mol. wt. = 836).

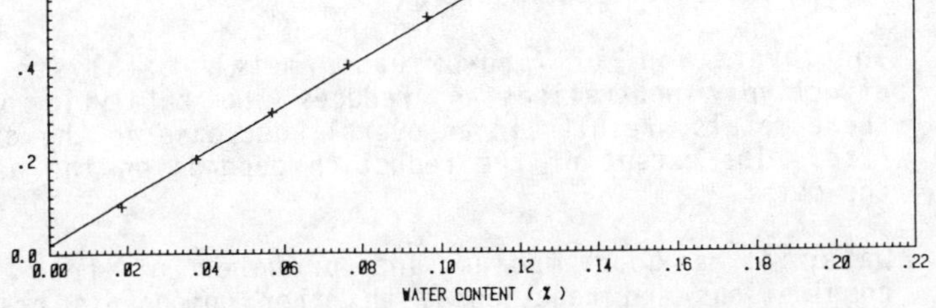

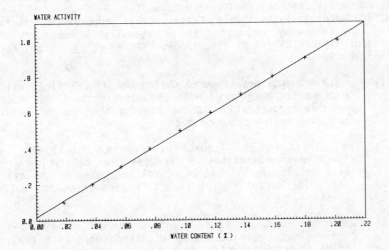

Figure 2. Water activity vs water content for crude palm oil at 50°C (PO mol. wt. = 836).

normally present in palm oil. However, the organic and inorganic substances that they contain may act as nutrients for the growth of microorganism in the oil, thus enhancing the rate of reaction. For chemical hydrolysis, although the rate is independent of the water concentration present in the oil where the water is in excess, the presence of the excess water above the solubility limit encourages the growth of lipolytic organisms. Hence under non-sterile conditions, the lipolytic rate is dependent on the dirt and water content of the oil to a certain extent. Below 50°C, therefore, enzymic hydrolysis occurs in conjunction with chemical hydrolysis. Above 50°C, mainly chemical hydrolysis occurs.

OXIDATIVE DETERIORATION

The effect of water on the oxidative deterioration of vegetable oil quality is an indirect phenomenon due to the fact that the reaction responsible for the oxidative deterioration does not involve the water moiety. The properties of water relevant to the oxidative stability of oils are related to the ability of water to :

(a) act as a solvent for dissolving chemical species
(b) act as a medium for the diffusion and reaction of chemical species within it and
(c) act as a species capable of diffusing and interacting with other species in various chemical ways.

In general, it was observed that as the water activity increases, the oxidative rate decreases. This decrease in oxidative reaction rate reaches a minimum at a certain water activity level and then increases again as the water activity increases. The level of water activity at which the deterioration rate is at a minimum depends on the nature of the system and varies from one food system to another.

At low water activity levels, Karel and his group of co-workers have observed the following mechanisms to be operatively important on the oxidation of lipids (Labuza _et al_., 1966 ; Maloney _et al_., 1966; Karel _et al_., 1967, 1974):-

(1) The hydroperoxides produced during the free radical oxidation reaction hydrogen bonds with the water present to lower the peroxide concentration, hence slowing the rate of initiation through peroxide decomposition.

(2) The hydration of pro-oxidant metal catalysts, which effectively neutralises or reduces the catalytic power of these metals, results in an overall decrease in the oxidation rate. The extent of the reduction depends on the nature of the metal.

(3) Water is a good medium and promoter of free radical combinations and reactions with other components present in lipids, thereby reducing the overall reaction rate.

(4) Water reacts with some metal catalysts to produce insoluble non-reactive hydroxides, thereby effectively removing the catalyst from the system (Kamiya _et al_., 1963).

At high a_w levels, the pro-oxidant effects of water could be attributed to :

(1) the increase in mobility of the hydrated metal catalysts to favourable reaction sites from their aqueous environment. The effect of this increase in mobility outweighs that of the hydration of the catalysts. This increase in mobility is due to a reduction in the viscosity of the system as was shown by Chou et al. (1973). This effect only partially accounts for the total increase in the oxidation rate at high water activity.

(2) the dissolution of precipitated catalysts, thus exposing new catalytic sites.

These above mentioned effects tend to result in the minimum-maximum type of oxidation curve as a function of the water activity as shown in Fig. 3. In the intermediate moisture range, the catalytic power of metals is an interplay between the concentration of the metals present (high or low) and the water content but not the water activity (Labuza et al., 1974). Water activity also affects the rate of photooxidation, but the reason(s) is/are not so well understood. Lipid-soluble antioxidants such as BHA, BHT and tocopherols are generally not affected by water content, but metal chelating agents are strongly influenced by changes in the water activity. The effectiveness of chelating agents such as EDTA and citric acid increases with increase in water activity, thus lowering the oxidation rate substantially. Verma & Prabhakar (1982) and Gopalakrishna (1983) have shown that the rate of peroxide formation is lower at higher water activities than at lower water activities for peanut oil and oxidising sunflower oil, respectively.

It is a well-accepted fact that there exists a range of a_w or moisture content whereby the presence of water would impart some degree of protection to a natural product or food system (Rockland & Nishi, 1980). This reduced rate of deterioration is ascribed to the activity of the water being able to modify the physical and chemical environment of the reacting species in the deteriorative reactions.

Chan (1979) attempted to define this water content range for crude palm oil at 40°C and 52°C via simulation of actual storage condition in vented storage bottles. His results for two months of storage supported earlier observations that the moisture level and temperature both affect the oxidative and hydrolytic reactions, with increasing temperature accelerating both reactions. However, moisture affects these two phenomena differently. Increasing moisture content (and hence a_w) accelerates the hydrolytic but inhibits the oxidative reactions as shown by the peroxide value curves in Fig. 4. His results indicate that there is a range of moisture content whereby the rate of deterioration is reduced. He advocated that increasing the moisture content in crude palm oil from 0.10% - 0.15% to 0.15% - 0.20% (corresponding to an a_w range of 0.51 - 0.75 to 0.75 - 1.00 at 50°) would result in insignificant increases in ffa during storage, but would minimise oxidative deterioration considerably.

A more detailed investigation of the effect of moisture levels on the quality deterioration of crude palm oil at 50°C was carried out by the present authors to extend the storage period studied by Chan (1979). Water was added to crude palm oil and then vacuumed to differing levels

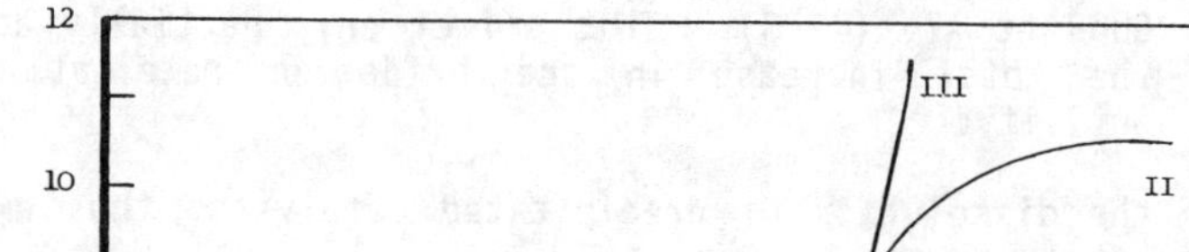

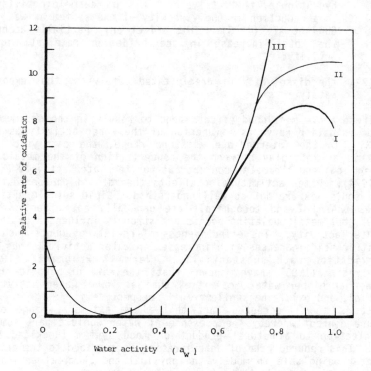

Figure 3. Overall effect of water activity on rates of lipid oxidation (Labuza, 1975).
Case I : dilution exceeds viscosity effect
Case II : dilution balances viscosity effect
Case III : viscosity effect predominates.

and stored in jars for about four months. Samples were withdrawn from the jars at periodic intervals for analysis via a septum and syringe arrangement in order to maintain the oil in a closed system. Moisture content was measured using a Baird and Tatlock automatic Karl Fisher titrator. Peroxide values and ffa were measured according to official AOCS methods. UV-vis measurements were carried out on 1% solutions of the oil in isooctane.

The moisture levels studied ranged from 0.06% to 0.26%. Fig. 5 shows the increase in ffa with storage time. As can be seen, the higher the moisture level, the higher is the ffa value with time and the least ffa increase occurred in the sample with the lowest moisture content (0.06%), except for the sample with 0.19% moisture which lies between the 0.06% and 0.10% curves. This observation is in line with the accepted concept of water being directly involved in the hydrolytic reaction. The anomaly observed in the sample with 0.19% moisture could be due to a reduction of water available for the hydrolytic reaction as a result of the protective mechanisms afforded by the water in the prevention of oxidative deterioration. From the figure, it can also be seen that there is not much difference between the acidification rates at the 0.21% and 0.26% moisture levels. This is to be expected since the solubility of water in palm oil at $50^{o}C$ is about 0.20%.

Primary and secondary oxidation characteristics are shown in Figs 6 and 7 in the forms of peroxide value and $UV_{269\ nm}$ absorbance-time curves, respectively. From Fig. 6, it can be seen that for the full storage period, the higher the moisture level, the slower is the rate of peroxide value increase, except for the 0.1% and 0.19% curves. The 0.1% moisture sample behaves in a manner similar to that of the 0.26% moisture sample while the 0.19% moisture sample behaves like the 0.21% moisture sample for about two months after which the rate of peroxide buildup is less than that of the 0.2% moisture sample. For storage periods of less than 50 days, the hierachy of the curves approximately follows that of the moisture levels, the highest moisture levels showing the most protective influence.

For the $UV_{269\ nm}$ absorbance curves, the situation is less well defined although the samples with the highest and lowest levels of $UV_{269\ nm}$ absorbance values were still the 0.06% and 0.26% moisture samples. The curves for the samples with moisture levels between 0.06% and 0.26% are bunched closely together, especially during the latter part of the storage period. The hierachy of the curves parallels that of the peroxide value curves to a large extent. This is to be expected since secondary oxidation products are derived from the breakdown of the hydroperoxides.

Figs 8-13 show the comparative oxidative and hydrolytic characteristics of the individual samples. At 0.06% moisture, there was only a small amount of hydrolytic activity, as indicated by the almost constant a_w level and the marginal increase in ffa, as opposed to the vast increases shown by the primary and secondary oxidation characteristics. At 0.10% moisture, there was a noticeable increase in the hydrolytic activity (an increase in ffa and a decrease in a_w with time), and an even more noticeable reduction in the oxidative activity. This reduction in oxidative activity could be due to a combination of the factors outlined earlier. A further increase in water content from 0.10% to 0.26% did not appear to affect the improvement in hydrolytic and oxidative stability to the same extent as that from 0.06% to 0.10%, except for the 0.19% moisture sample. In the case of the 0.19% moisture sample, the primary oxidation level was approximately that of the 0.10% moisture sample, but the ffa level was lower. A moisture level of 0.19% appears to be optimal for palm oil, as evidenced by the minimisation of the acidification rate and the maximisation of the protective influence of moisture with respect to oxidation. The results support the observation by Chan (1979) that there exists an optimal range of moisture for crude palm oil in the range of 0.15% to 0.20% with the optimal level

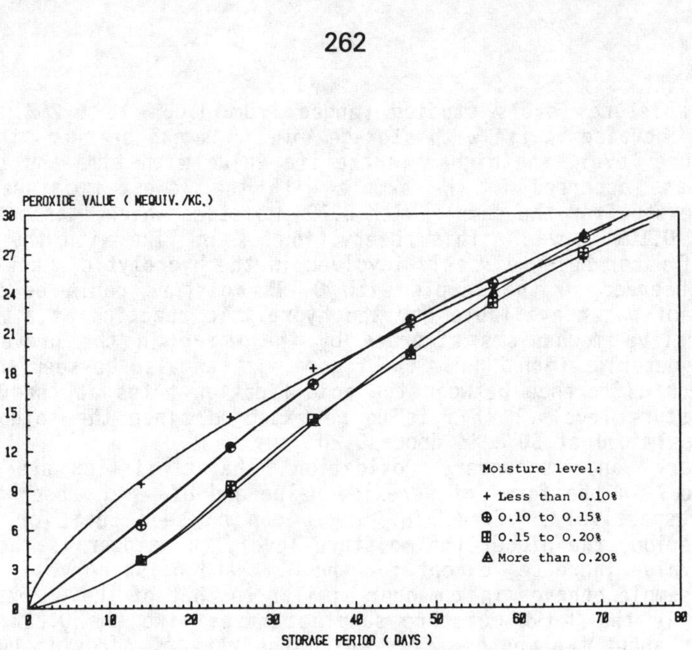

Figure 4. PV storage curves of CPO at 50°C (from Chan, 1979).

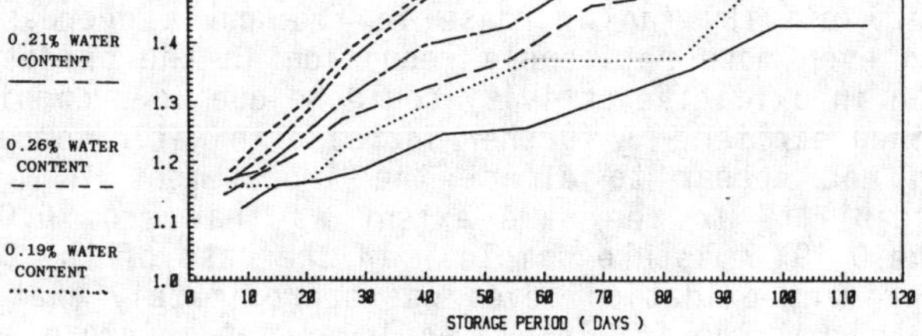

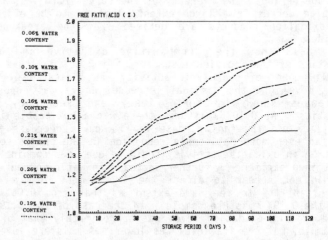

Figure 5. Changes in FFA of CPO stored at 50°C.

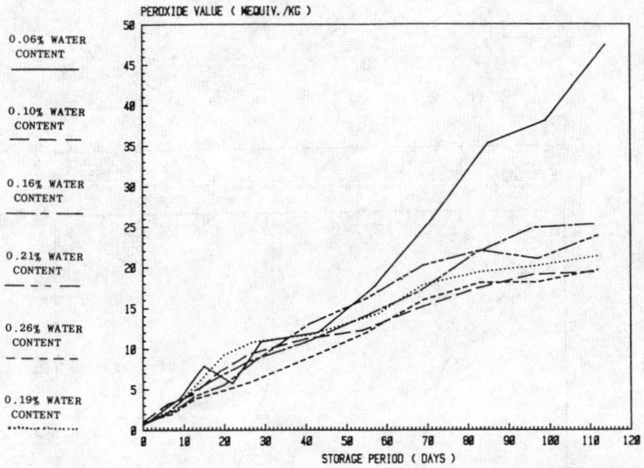

Figure 6. Changes in PV of CPO stored at 50°C.

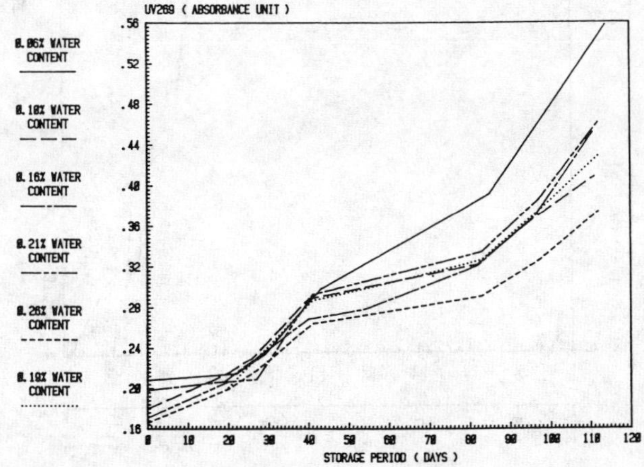

Figure 7. Changes in UV_{269} absorbance of CPO stored at 50°C.

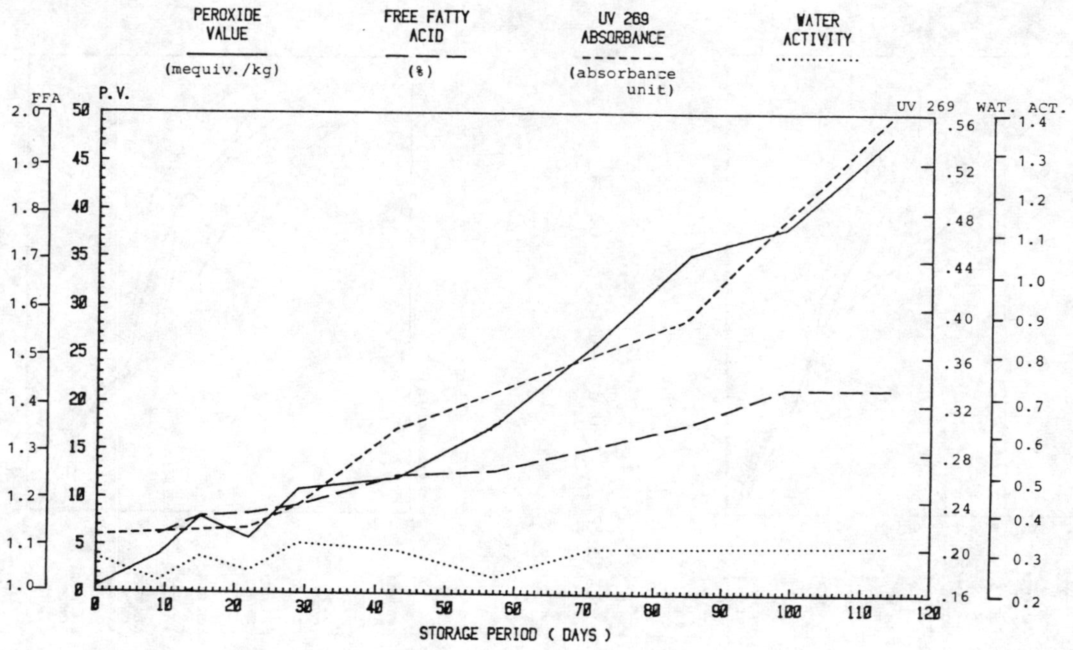

Figure 8. Characteristics of 0.06% moisture sample stored at 50°C.

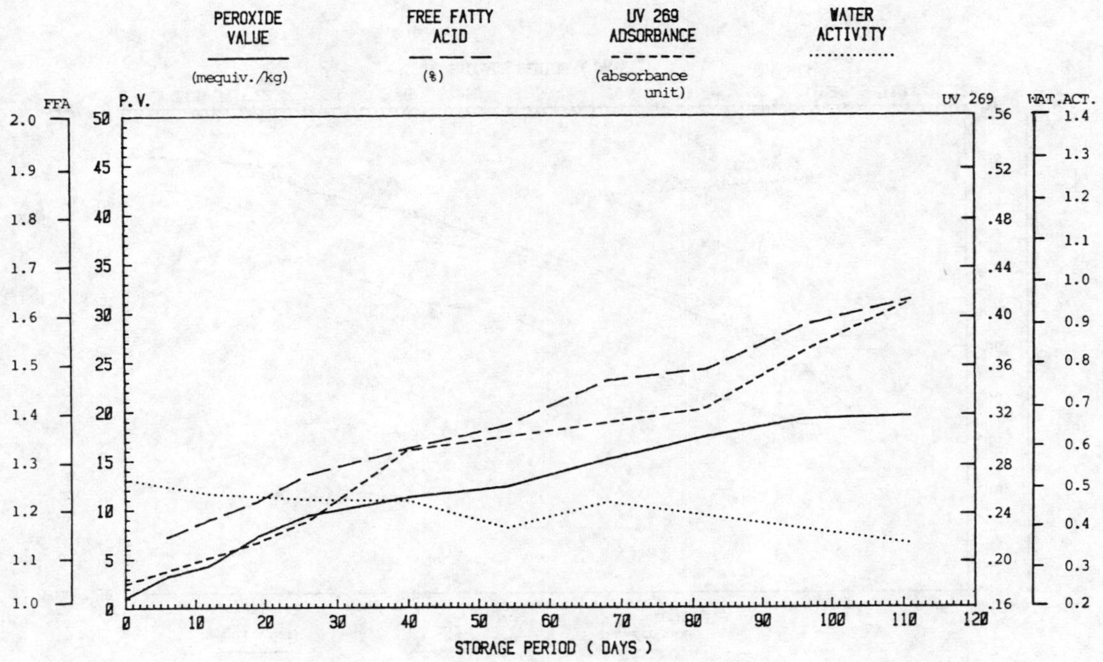

Figure 9. Characteristics of 0.16% moisture sample stored at 50°C.

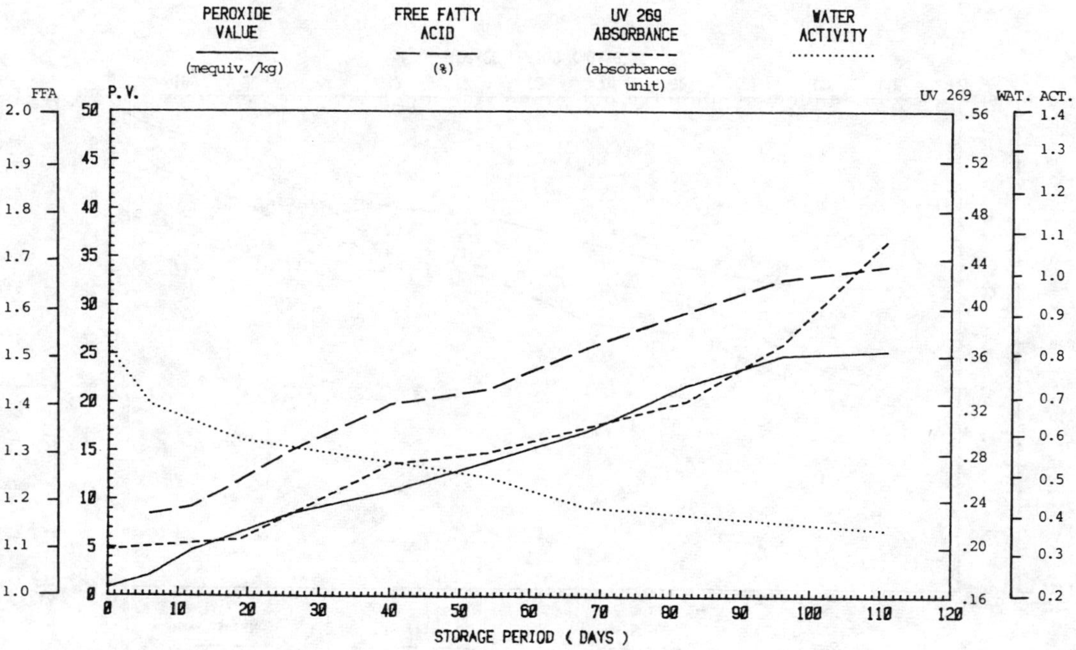

Figure 10. Characteristics of 0.16% moisture sample stored at 50°C.

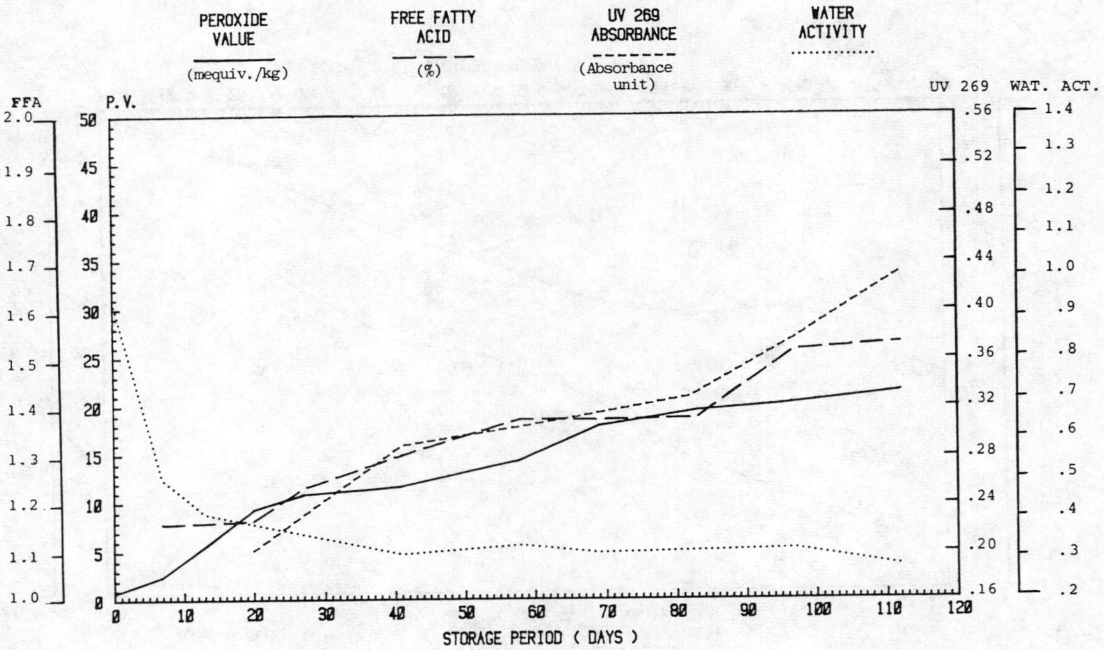

Figure 11. Characteristics of 0.19% moisture sample stored at 50°C.

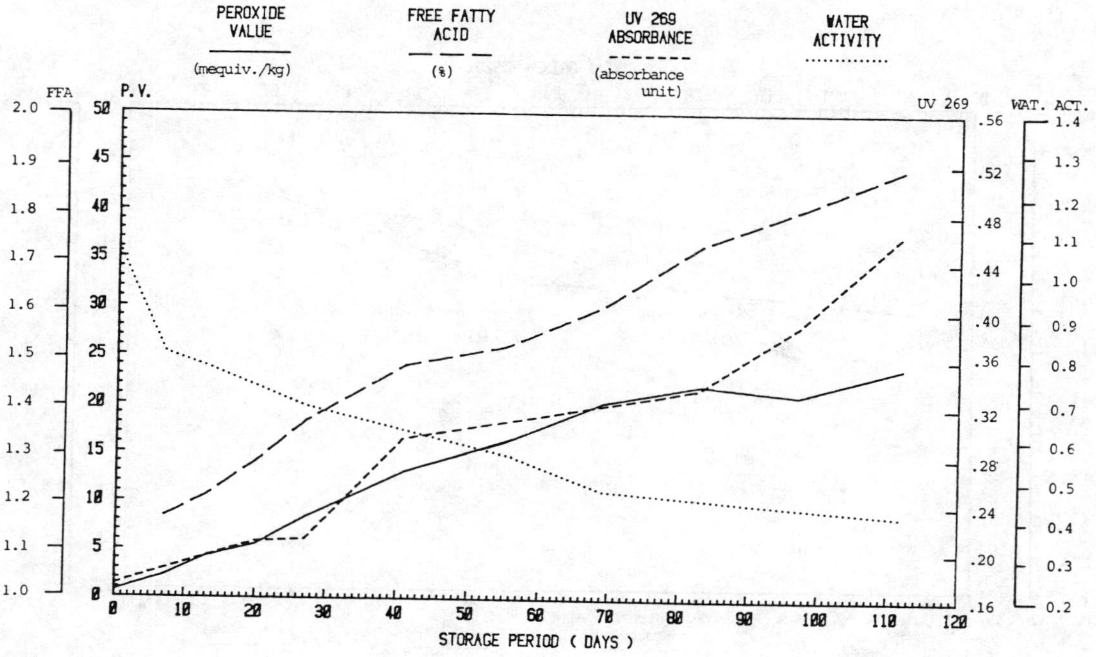

Figure 12. Characteristics of 0.21% moisture sample stored at 50°C.

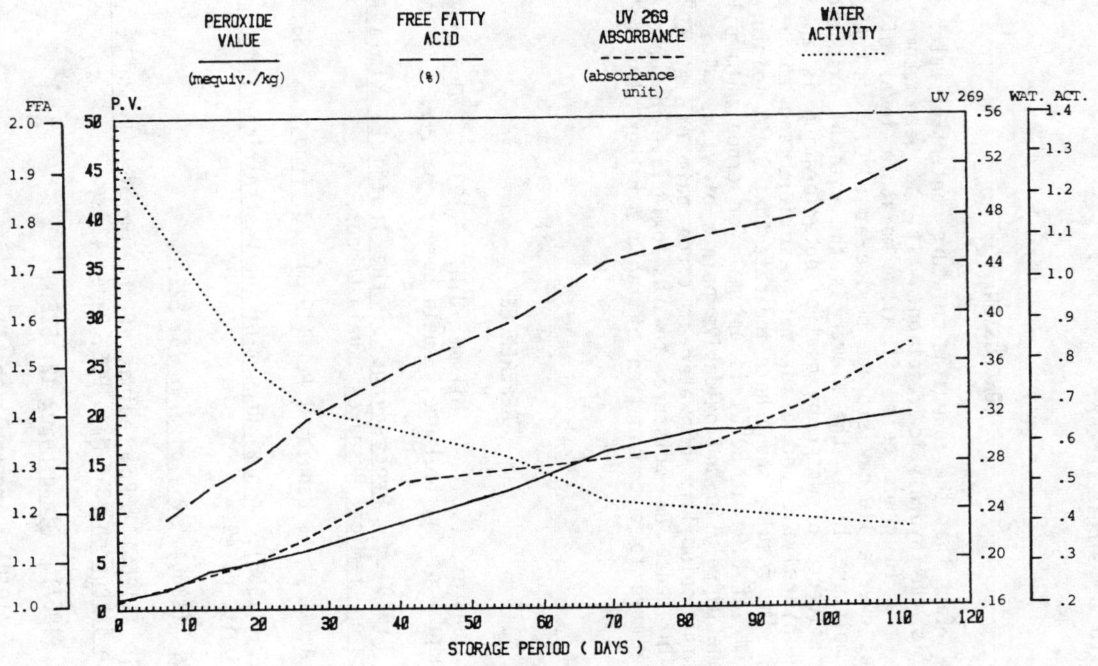

Figure 13. Characteristics of 0.26% moisture sample stored at 50 °C.

more towards the higher end. From both the hydrolytic and oxidative points of view, a moisture content of 0.19% (corresponding to an a_w of about 0.9) appears to be ideal for crude palm oil storage at 50°. This high a_w tends to imply that the dissolved water is more in the free rather than in the bound form. Hence the protective influence on the oil is probably due more to a combination of the factors outlined earlier rather than to any single factor.

CONCLUSION

It is a known fact that the lower the moisture content or water activity, the slower is the hydrolytic reaction. It has been shown that the increase in ffa in crude palm oil at a moisture level of 0.06% is insignificant over a 110-day storage period at 50^oC. In contrast, the higher the moisture level, the slower is the rate of oxidation. The protective influence of water on quality deterioration is an interplay between the hydrolytic and oxidative reactions and their chemical environment. The results shown here indicate that 0.19% moisture (with a water activity of 0.94 at 50^oC) is optimal for crude palm oil storage from both the hydrolytic and oxidative points of view. This level is well below the solubility of water in crude palm oil at 50^oC. The protective influence of moisture towards oxidative deterioration is probably due more to a combination of factors rather than to any one factor.

REFERENCES

Bunting, B., Georgi, C.D.V. & Milsum, J.N. (1934). The Oil Palm in Malaya Dept. of Agriculture, Kuala Lumpur, Malaya.

Chan, K.S. (1979). Proceedings of the Institute Kimia Malaysia Conference on Chemical Research, Kuala Lumpur.

Chou, H.E., Acott, K. & Labuza, T.P. (1973). J. Food Sci. **38**, 316.

Coursey, D.G. (1958). Annual Report of the West African Stored Products Research Unit, p. 49.

Coursey, D.G. (1960). Oleagineaux **15**, 623.

Coursey, D.G. (1965). Proceedings of the Tropical Products Institute Conference in London, 3-6 May 1965, Paper 4.

Desassi, A. (1957). Oleagineaux **12**, 525.

Fickendey, E. (1910). Tropenplanzer **14**, 560.

Gopalakrishna, A.G. & Prabhakar, J.V. (1983). JAOCS **60**(5), 968.

Heidelbaugh, N.D. & Karel, M. (1970). JAOCS **47**(12), 541.

Hilder, H. (1968). JAOCS **45**(10), 703.

Kamiya, Y., Beaton, S., Lafortune, S. & Ingold, K.V. (1963). Can. J. Chem. **41**, 2034.

Karel, M., Labuza, T.P. & Maloney, J.F. (1967). Cryobiology **3**, 1288.

Karel, M. (1974). Food Technol. **28**, 50.

Labuza, T.P., Maloney, J.F. & Karel, M. (1966). J. Food Sci. **31**, 885.

Labuza, T.P., Tannenbaum, S.R. & Karel, M. (1970). Food Technol. **24**, 543.

Labuza, T.P. McNally, L., Gallager, D., Hawkes, J. & Hurtado, F. (1972). J. Food Sci. **37**, 154.

Labuza, T.P. & Chan, H.E. (1974). J. Food Sci. **39**, 112.

Labuza, T.P. (1975). In Water Relations of Food (Duckworth, R.B., ed.), p. 455. Academic Press, London.

Loncin, M. (1952). Thesis on "L' Hydrolyse spontanse' des Huiles Glyceridiques et en particulier de l'Huile de Palme". C.E.R.I.A., Brussels, Belgium.

Loncin, M. (1955). Fette und Seifen **41**, 413.

Loncin, M. (1956). Rev. Fran. des. Corps. Gras **3**, 255.

Maloney, J., Labuza, T.P., Wallace, D.H. & Karel, M. (1966). J. Food Sci. **31**, 878.

Quast, D.G. & Karel, M. (1971). J. Food Technol. **6**, 96.

Rockland, L.B. & Nishi, S.K. (1980). Food Technol. **34**(4), 42.

Stork Palm Oil Review (1960) Vol. **1**, No. 1. (February).

Vanneck, C., Loncin, M. & Jacqmain, D. (1951). Bull. Agric du Congo Belge. **42**, 57.

Verma, M.M. & Prabhakar, J.V. (1982). Ind. Food Packer **36**(5), 77.

Vervloet, C. (1953). De Bergcultures **22**, 218.

SUBJECT INDEX

A

B

C

D

H

I

J, K, L

M

N

O

P

Q, R

S

T, U, V

W, X, Y, Z